SpringerBriefs in Applied Sciences and Technology

For further volumes:
http://www.springer.com/series/8884

Tarek I. Zohdi

Dynamics of Charged Particulate Systems

Modeling, Theory and Computation

Springer

Prof. Tarek I. Zohdi
Department of Mechanical Engineering
University of California, Berkeley
Etcheverry Hall 6195
Berkeley
CA 94720-1740
USA

ISSN 2191-530X
ISBN 978-3-642-28518-9
DOI 10.1007/978-3-642-28519-6
Springer Heidelberg New York Dordrecht London

e-ISSN 2191-5318
e-ISBN 978-3-642-28519-6

Library of Congress Control Number: 2012933323

Printed on acid-free paper

Springer is part of Springer Science+Business Media (www.springer.com)

Dedicated to my mother,
Omnia El-Menshawy, who supported
me throughout my life.
All profits generated by this book will be
donated to the United Nations Children's
Fund (UNICEF).

 I am certain that, despite painstaking efforts, there remain errors of one sort or another in this monograph. Therefore, I would be grateful if readers who find such flaws would contact me at zohdi@me.berkeley.edu.

Preface

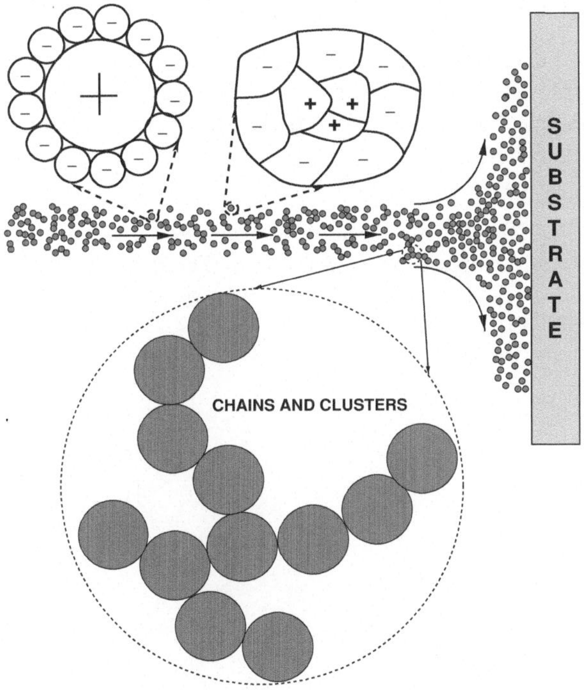

Fig. 1 Spray processing of a surface with a charged particulate spray. Applications can vary from (1) epitaxy (building particulate layers), (2) particulate implantation/infiltration and (3) surface ablation (Zohdi [1])

The objective of this monograph is to provide a concise introduction to the dynamics of systems comprised of charged small-scale particles. Flowing, small-scale, particles ("particulates") are ubiquitous in industrial processes and in the natural sciences. Applications include electrostatic copiers, inkjet printers, powder coating machines, etc., and a variety of manufacturing processes. Due to their small-scale size, external electromagnetic fields can be utilized to manipulate and control charged particulates in industrial processes in order to achieve results that are not possible by purely mechanical means alone. A unique feature of small-scale particulate flows is that they exhibit a strong sensitivity to interparticle near-field forces, leading to nonstandard particulate dynamics, agglomeration and cluster formation, which can strongly affect manufactured product quality. This monograph also provides an introduction to the mathematically-related topic of the dynamics of swarms of interacting objects, which has gained the attention of a number of scientific communities. In summary, the following topics are discussed in detail:

1. Dynamics of an individual charged particle,
2. Dynamics of rigid clusters of charged particles,
3. Dynamics of flowing charged particles,
4. Dynamics of charged particle impact with electrified surfaces and
5. An introduction to mechanistic modeling of swarms.

The text can be viewed as a research monograph suitable for use in an upper division undergraduate or first year graduate course geared towards students in the applied sciences, mechanics and mathematics that have an interest in the analysis of particulate materials.

1. Zohdi, T. I. (2010). On the dynamics of charged electromagnetic particulate jets. *Archives of Computational Methods in Engineering*, *17*(2), 109–135.

Berkeley, January 2012 T. I. Zohdi

Contents

Chapter 1
Introduction: Dynamics of an Individual Charged Particle

We start with an introduction to the dynamics of a single charged particle, and then progress to rigid clusters, and then flowing systems.

1.1 Notation

In this work, boldface symbols imply vectors or tensors. *A fixed Cartesian coordinate system will be used throughout this monograph.* The unit vectors for such a system are given by the mutually orthogonal triad $(\boldsymbol{e}_1, \boldsymbol{e}_2, \boldsymbol{e}_3)$. For the inner product of two vectors, $\boldsymbol{u}$ and $\boldsymbol{v}$, in three dimensions, we have

$$\boldsymbol{u} \cdot \boldsymbol{v} = \sum_{i=1}^{3} u_i v_i = u_1 v_1 + u_2 v_2 + u_3 v_3 = ||\boldsymbol{u}|| ||\boldsymbol{v}|| cos\theta, \tag{1.1}$$

where

$$||\boldsymbol{u}|| = \sqrt{u_1^2 + u_2^2 + u_3^2} \tag{1.2}$$

represents the Euclidean norm in R^3 and θ is the angle between them. The cross (vector) product of two vectors is

$$\boldsymbol{u} \times \boldsymbol{v} = -\boldsymbol{v} \times \boldsymbol{u} = \begin{vmatrix} \boldsymbol{e}_1 & \boldsymbol{e}_2 & \boldsymbol{e}_3 \\ u_1 & u_2 & u_3 \\ v_1 & v_2 & v_3 \end{vmatrix} = ||\boldsymbol{u}|| ||\boldsymbol{v}|| sin\theta \, \boldsymbol{n}, \tag{1.3}$$

where $\boldsymbol{n}$ is the unit normal to the plane formed by the vectors $\boldsymbol{u}$ and $\boldsymbol{v}$. The temporal differentiation of a vector is given by

T. I. Zohdi, *Dynamics of Charged Particulate Systems*, SpringerBriefs in Applied Sciences and Technology, DOI: 10.1007/978-3-642-28519-6_1,

$$\frac{d}{dt}\boldsymbol{u} = \frac{du_1}{dt}\boldsymbol{e}_1 + \frac{du_2}{dt}\boldsymbol{e}_2 + \frac{du_3}{dt}\boldsymbol{e}_3 = \dot{u}_1\boldsymbol{e}_1 + \dot{u}_2\boldsymbol{e}_2 + \dot{u}_3\boldsymbol{e}_3. \tag{1.4}$$

1.2 Kinematics of a Single Particle

We denote the position of a point in space by the vector $\boldsymbol{r}$. The instantaneous velocity of a point is given by

$$\boldsymbol{v} = \frac{d\boldsymbol{r}}{dt} = \dot{\boldsymbol{r}}. \tag{1.5}$$

The instantaneous acceleration of a point is given by

$$\boldsymbol{a} = \frac{d\boldsymbol{v}}{dt} = \dot{\boldsymbol{v}} = \ddot{\boldsymbol{r}}. \tag{1.6}$$

In fixed Cartesian coordinates, for position, we have

$$\boldsymbol{r} = r_1\boldsymbol{e}_1 + r_2\boldsymbol{e}_2 + r_3\boldsymbol{e}_3, \tag{1.7}$$

for velocity

$$\boldsymbol{v} = \dot{\boldsymbol{r}} = \dot{r}_1\boldsymbol{e}_1 + \dot{r}_2\boldsymbol{e}_2 + \dot{r}_3\boldsymbol{e}_3, \tag{1.8}$$

and for acceleration

$$\boldsymbol{a} = \ddot{\boldsymbol{r}} = \ddot{r}_1\boldsymbol{e}_1 + \ddot{r}_2\boldsymbol{e}_2 + \ddot{r}_3\boldsymbol{e}_3. \tag{1.9}$$

Their magnitudes are denoted by $||\boldsymbol{r}|| = \sqrt{\boldsymbol{r}\cdot\boldsymbol{r}}$, $||\boldsymbol{v}|| = \sqrt{\boldsymbol{v}\cdot\boldsymbol{v}}$ and $||\boldsymbol{a}|| = \sqrt{\boldsymbol{a}\cdot\boldsymbol{a}}$. The relative position vector between a point i with respect to a point j is denoted by $\boldsymbol{r}_{i-j} = \boldsymbol{r}_i - \boldsymbol{r}_j$, the relative velocity vector by $\boldsymbol{v}_{i-j} = \boldsymbol{v}_i - \boldsymbol{v}_j$ and the relative acceleration vector by $\boldsymbol{a}_{i-j} = \boldsymbol{a}_i - \boldsymbol{a}_j$. Also, we denote $\boldsymbol{r}_{i\to j} = \boldsymbol{r}_j - \boldsymbol{r}_i$, $\boldsymbol{v}_{i\to j} = \boldsymbol{v}_j - \boldsymbol{v}_i$ and $\boldsymbol{a}_{i\to j} = \boldsymbol{a}_j - \boldsymbol{a}_i$.

1.3 Kinetics of a Single Particle

Throughout this monograph, the fundamental relation between force and acceleration is given by Newton's second law of motion, in vector form:

$$\boldsymbol{\Psi} = \frac{d}{dt}(m\boldsymbol{v}) \tag{1.10}$$

where $\boldsymbol{\Psi}$ is the sum (resultant) of all the applied forces instantaneously acting on mass m.

1.3.1 Impulse and Momentum

Newton's second law can be rewritten as

$$\boldsymbol{\Psi} = \frac{d(m\boldsymbol{v})}{dt} \Rightarrow \boldsymbol{G}(t_1) + \int_{t_1}^{t_2} \boldsymbol{\Psi}\, dt = \boldsymbol{G}(t_2), \tag{1.11}$$

where

$$\boldsymbol{G}(t_1) = (m\boldsymbol{v})|_{t=t_1} \tag{1.12}$$

is the linear momentum. Clearly, if $\boldsymbol{\Psi} = \boldsymbol{0}$, then $\boldsymbol{G}(t_1) = \boldsymbol{G}(t_2)$, and linear momentum is said to be conserved. A related quantity is the angular momentum. About the origin

$$\boldsymbol{H}_o \stackrel{\text{def}}{=} \boldsymbol{r} \times m\boldsymbol{v}. \tag{1.13}$$

Clearly, the moment $\boldsymbol{M}$ implies

$$\boldsymbol{M} = \boldsymbol{r} \times \boldsymbol{\Psi} = \frac{d\boldsymbol{H}_o}{dt} \Rightarrow \boldsymbol{H}_o(t_1) + \int_{t_1}^{t_2} \underbrace{\boldsymbol{r} \times \boldsymbol{\Psi}}_{\boldsymbol{M}}\, dt = \boldsymbol{H}_o(t_2). \tag{1.14}$$

Thus, if $\boldsymbol{M} = \boldsymbol{0}$, then $\boldsymbol{H}_o(t_1) = \boldsymbol{H}_o(t_2)$, and angular momentum is said to be conserved.

1.4 Dynamics in the Presence of an Electromagnetic Field

We recall the following important observations in conjunction with electromagnetic phenomena [24]:

- If a point charge q experiences a force $\boldsymbol{\Psi}^e$, the electric field, $\boldsymbol{E}$, at the location of the charge is defined by $\boldsymbol{\Psi}^e = q\boldsymbol{E}$.
- If the charge is moving, another force may arise, $\boldsymbol{\Psi}^m$, which is proportional to its velocity $\boldsymbol{v}$. This other (induced) field is denoted as the "magnetic induction" or just the "magnetic field," $\boldsymbol{B}$, such that $\boldsymbol{\Psi}^m = q\boldsymbol{v} \times \boldsymbol{B}$.
- If the forces occur concurrently (the charge is moving through the region possessing both electric and magnetic fields), then the electromagnetic force is $\boldsymbol{\Psi}^{em} = q\boldsymbol{E} + q\boldsymbol{v} \times \boldsymbol{B}$.

We consider an isolated charged mass with position vector denoted by $\boldsymbol{r}$, governed by ($\dot{\boldsymbol{r}} = \boldsymbol{v}$, $\ddot{\boldsymbol{r}} = \dot{\boldsymbol{v}}$)

$$m\dot{\boldsymbol{v}} = q(\boldsymbol{E} + \boldsymbol{v} \times \boldsymbol{B}). \tag{1.15}$$

The governing Eq. 1.15, written in component form is, for component 1

$$\dot{v}_1 = \frac{q}{m}(E_1 + (v_2 B_3 - v_3 B_2)), \tag{1.16}$$

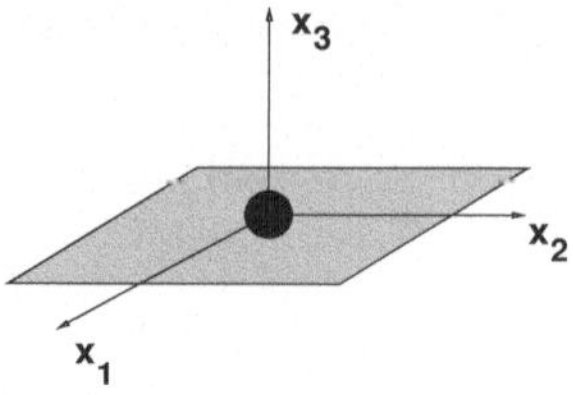

Fig. 1.1 An isolated particle with an applied "dead" electromagnetic field for special cases # 1 to # 3

for component 2

$$\dot{v}_2 = \frac{q}{m}(E_2 - (v_1 B_3 - v_3 B_1)), \tag{1.17}$$

and for component 3

$$\dot{v}_3 = \frac{q}{m}(E_3 + (v_1 B_2 - v_2 B_1)). \tag{1.18}$$

With the appropriate simplifications, for example, that $\boldsymbol{E}^{ext} = (E_1^{ext}, E_2^{ext}, E_3^{ext})$ is an independent (not dependent on the particles) external electric field and $\boldsymbol{B}^{ext} = (B_1^{ext}, B_2^{ext}, B_3^{ext})$ is an independent external magnetic field, the equations can be solved analytically (for example, see Jackson [24]) for an isolated particle (Fig. 1.1). We consider a few of these cases, for static ("dead") electromagnetic fields.

1.4.1 Special Case # 1: No Magnetic Field ($\boldsymbol{E}^{ext} = E_3^{ext}\boldsymbol{e}_3$ and $\boldsymbol{B}^{ext} = \boldsymbol{0}$)

In the special case when there is no magnetic field, if $\boldsymbol{r}(t = 0) = \boldsymbol{0}$, $\boldsymbol{v}(t = 0) = v_o\boldsymbol{e}_1$, $\boldsymbol{B}^{ext} = \boldsymbol{0}$ and $\boldsymbol{E}^{ext} = E_3^{ext}\boldsymbol{e}_3$, the solution for the dynamics of an isolated particle is

$$\left\{\begin{matrix} v_1(t) \\ v_2(t) \\ v_3(t) \end{matrix}\right\} = \left\{\begin{matrix} v_o \\ 0 \\ \frac{q}{m}E_3^{ext}t \end{matrix}\right\} \Rightarrow \left\{\begin{matrix} r_1(t) \\ r_2(t) \\ r_3(t) \end{matrix}\right\} = \left\{\begin{matrix} v_o t \\ 0 \\ \frac{q}{2m}E_3^{ext}t^2 \end{matrix}\right\}. \tag{1.19}$$

Thus, the third position component grows quadratically in time.

1.4.2 Special Case # 2: No Electric Field ($\boldsymbol{E}^{ext} = \boldsymbol{0}$ and $\boldsymbol{B}^{ext} = B_3^{ext}\boldsymbol{e}_3$)

Now, consider a case with no electric field and a magnetic field present, $\boldsymbol{r}(t = 0) = \boldsymbol{0}$, $\boldsymbol{v}(t = 0) = v_o\boldsymbol{e}_1$, $\boldsymbol{B}^{ext} = B_3^{ext}\boldsymbol{e}_3$ and $\boldsymbol{E}^{ext} = \boldsymbol{0}$. Consequently, for a single particle, the solution is

$$\begin{Bmatrix} v_1(t) \\ v_2(t) \\ v_3(t) \end{Bmatrix} = \begin{Bmatrix} v_o cos\omega t \\ -v_o sin\omega t \\ 0 \end{Bmatrix} \Rightarrow \begin{Bmatrix} r_1(t) \\ r_2(t) \\ r_3(t) \end{Bmatrix} = \begin{Bmatrix} \frac{v_o}{\omega} sin\omega t \\ \frac{v_o}{\omega}(cos\omega t - 1) \\ 0 \end{Bmatrix}, \tag{1.20}$$

where $\omega = \frac{qB_3^{ext}}{m}$ is known as the cyclotron frequency. The cyclotron frequency (gyrofrequency) is the angular frequency at which a charged particle makes circular orbits in a plane perpendicular to the static magnetic field. Notice that when $E_3^{ext} = 0$, this traces out the equation of a circle centered at $(0, \frac{-v_o}{\omega}, 0)$. The radius of the "magnetically-induced circle" (radius of oscillation) is[1]

$$R \stackrel{\text{def}}{=} \frac{v_o}{\omega} = \frac{v_o m}{qB_3^{ext}}. \tag{1.21}$$

Thus, if a desired "turning radius" is denoted by R, one may solve for the magnetic field that delivers the desired effect, $B_3^{ext} = \frac{v_o m}{qR}$. We define the corresponding time period for one cycle to be completed as $T \stackrel{\text{def}}{=} 2\pi/\omega$.

1.4.3 Special Case # 3: Combined Electric and Magnetic Fields ($\boldsymbol{E}^{ext} = E_3^{ext}\boldsymbol{e}_3$ and $\boldsymbol{B}^{ext} = B_3^{ext}\boldsymbol{e}_3$)

Now, consider both the electric and magnetic fields to be present, $\boldsymbol{r}(t = 0) = \boldsymbol{0}$, $\boldsymbol{v}(t = 0) = v_o\boldsymbol{e}_1$, $\boldsymbol{B}^{ext} = B_3^{ext}\boldsymbol{e}_3$ and $\boldsymbol{E}^{ext} = E_3^{ext}\boldsymbol{e}_3$, consequently, for a single particle

$$\begin{Bmatrix} v_1(t) \\ v_2(t) \\ v_3(t) \end{Bmatrix} = \begin{Bmatrix} v_o cos\omega t \\ -v_o sin\omega t \\ \frac{q}{m}E_3^{ext}t \end{Bmatrix} \Rightarrow \begin{Bmatrix} r_1(t) \\ r_2(t) \\ r_3(t) \end{Bmatrix} = \begin{Bmatrix} \frac{v_o}{\omega} sin\omega t \\ \frac{v_o}{\omega}(cos\omega t - 1) \\ \frac{q}{2m}E_3^{ext}t^2 \end{Bmatrix}. \tag{1.22}$$

Thus, the third position component grows quadratically in time, while the other two components trace out a circle. The composite motion is that of a helix.

1.4.4 General Solutions: Magnetic Rotation Axes

The term $\boldsymbol{v} \times \boldsymbol{B}^{ext}$ can explicitly be written in the following matrix form:

$$\boldsymbol{v} \times \boldsymbol{B}^{ext} = \begin{bmatrix} 0 & B_3^{ext} & -B_2^{ext} \\ -B_3^{ext} & 0 & B_1^{ext} \\ B_2^{ext} & -B_1^{ext} & 0 \end{bmatrix} \begin{Bmatrix} v_1 \\ v_2 \\ v_3 \end{Bmatrix} \stackrel{\text{def}}{=} \boldsymbol{A}_B \cdot \boldsymbol{v}. \tag{1.23}$$

[1] This field generates helical motion in three dimensions when $\boldsymbol{E}^{ext} \neq \boldsymbol{0}$.

The system of equations can be written as

$$\dot{\boldsymbol{v}} - \frac{q}{m}\boldsymbol{A}_B \cdot \boldsymbol{v} = \frac{q}{m}\boldsymbol{E}. \tag{1.24}$$

We define an "eigensystem" via

$$\boldsymbol{v} = \boldsymbol{T} \cdot \hat{\boldsymbol{v}} \tag{1.25}$$

and insert into the governing equations to yield

$$\dot{\hat{\boldsymbol{v}}} - \frac{q}{m}(\boldsymbol{T}^{-1} \cdot \boldsymbol{A}_B \cdot \boldsymbol{T}) \cdot \hat{\boldsymbol{v}} = \frac{q}{m}\boldsymbol{T}^{-1} \cdot \boldsymbol{E}. \tag{1.26}$$

The proper choice of $\boldsymbol{T}$ to decouple the system is to form $\boldsymbol{T}$ from the eigenvector of $\boldsymbol{A}_B$. The eigenvalues are computed from

$$|\boldsymbol{A}_B - \lambda \mathbf{1}| = 0 \Rightarrow \begin{vmatrix} -\lambda & B_3^{ext} & -B_2^{ext} \\ -B_3^{ext} & -\lambda & B_1^{ext} \\ B_2^{ext} & -B_1^{ext} & -\lambda \end{vmatrix} = 0, \tag{1.27}$$

which implies $\lambda_1 = 0$, $\lambda_2 = -i||\boldsymbol{B}^{ext}||$ and $\lambda_3 = +i||\boldsymbol{B}^{ext}||$. The first eigenvector, corresponding to $\lambda_1 = 0$, is

$$\boldsymbol{\Lambda}^{(1)} = \frac{1}{||\boldsymbol{B}^{ext}||} \begin{Bmatrix} B_1^{ext} \\ B_2^{ext} \\ B_3^{ext} \end{Bmatrix} \tag{1.28}$$

The second eigenvector, corresponding to $\lambda_2 = -i||\boldsymbol{B}^{ext}||$, is

$$\boldsymbol{\Lambda}^{(2)} = \frac{1}{\gamma} \begin{Bmatrix} B_1^{ext} B_2^{ext} - iB_3^{ext}||\boldsymbol{B}^{ext}|| \\ -(B_1^{ext})^2 - (B_3^{ext})^2 \\ B_2^{ext} B_3^{ext} + iB_1^{ext}||\boldsymbol{B}^{ext}|| \end{Bmatrix} \tag{1.29}$$

where

$$\gamma = \sqrt{|(B_1^{ext})^2 + (B_3^{ext})^2|^2 + |iB_1^{ext}||\boldsymbol{B}^{ext}|| + B_2^{ext}B_3^{ext}|^2 + |B_1^{ext}B_2^{ext} - iB_3^{ext}||\boldsymbol{B}^{ext}|||^2}. \tag{1.30}$$

The second eigenvector, corresponding to $\lambda_2 = i||\boldsymbol{B}^{ext}||$, is

$$\boldsymbol{\Lambda}^{(3)} = \frac{1}{\hat{\gamma}} \begin{Bmatrix} B_1^{ext} B_2^{ext} + iB_3^{ext}||\boldsymbol{B}^{ext}|| \\ -(B_1^{ext})^2 - (B_3^{ext})^2 \\ B_2^{ext} B_3^{ext} - iB_1^{ext}||\boldsymbol{B}^{ext}|| \end{Bmatrix} \tag{1.31}$$

where

$$\hat{\gamma} = \sqrt{|-(B_1^{ext})^2 - (B_3^{ext})^2|^2 + |B_1^{ext}B_2^{ext} + iB_3^{ext}||\boldsymbol{B}^{ext}|||^2 + |B_2^{ext}B_3^{ext} - iB_1^{ext}||\boldsymbol{B}^{ext}|||^2}. \tag{1.32}$$

The first eigenvector indicates the direction of the axis of rotation, while the second and third (conjugate) eigenvalues dictate the cyclotron frequency and the radius of the helical circle.

Remark 1 One can decouple the coupled (vector component) system by forming a matrix from the following set of eigenvectors:

$$\left\{\begin{matrix} \Lambda_1^{(1)} \\ \Lambda_2^{(1)} \\ \Lambda_3^{(1)} \end{matrix}\right\}_{\lambda_1}, \left\{\begin{matrix} \Lambda_1^{(2)} \\ \Lambda_2^{(2)} \\ \Lambda_3^{(2)} \end{matrix}\right\}_{\lambda_2}, \left\{\begin{matrix} \Lambda_1^{(3)} \\ \Lambda_2^{(3)} \\ \Lambda_3^{(3)} \end{matrix}\right\}_{\lambda_3}. \tag{1.33}$$

Performing a similarity transform to decouple the system, we obtain

$$\begin{aligned} &\left\{\begin{matrix} \dot{\hat{v}}_1 \\ \dot{\hat{v}}_2 \\ \dot{\hat{v}}_3 \end{matrix}\right\} - \frac{qi}{m}\begin{bmatrix} 0 & 0 & 0 \\ 0 & ||\boldsymbol{B}^{ext}|| & 0 \\ 0 & 0 & -||\boldsymbol{B}^{ext}|| \end{bmatrix}\left\{\begin{matrix} \hat{v}_1 \\ \hat{v}_2 \\ \hat{v}_3 \end{matrix}\right\} \\ &= \frac{q}{m}\begin{bmatrix} \Lambda_1^{(1)} & \Lambda_1^{(2)} & \Lambda_1^{(3)} \\ \Lambda_2^{(1)} & \Lambda_2^{(2)} & \Lambda_2^{(3)} \\ \Lambda_3^{(1)} & \Lambda_3^{(2)} & \Lambda_3^{(3)} \end{bmatrix}^{-1}\left\{\begin{matrix} E_1^{ext} \\ E_2^{ext} \\ E_3^{ext} \end{matrix}\right\} \stackrel{\text{def}}{=} \left\{\begin{matrix} \hat{f}_1 \\ \hat{f}_2 \\ \hat{f}_3 \end{matrix}\right\}. \end{aligned} \tag{1.34}$$

The decoupled problems can be written as

$$\dot{\hat{v}}_1 = \hat{f}_1, \tag{1.35}$$

$$\dot{\hat{v}}_2 - i\frac{q}{m}||\boldsymbol{B}^{ext}||\hat{v}_2 = \hat{f}_2, \tag{1.36}$$

$$\dot{\hat{v}}_3 + i\frac{q}{m}||\boldsymbol{B}^{ext}||\hat{v}_3 = \hat{f}_3, \tag{1.37}$$

and can be solved individually (decoupled). Afterwards, the solution in the transformed space is transformed back to yield

$$\begin{bmatrix} \Lambda_1^{(1)} & \Lambda_1^{(2)} & \Lambda_1^{(3)} \\ \Lambda_2^{(1)} & \Lambda_2^{(2)} & \Lambda_2^{(3)} \\ \Lambda_3^{(1)} & \Lambda_3^{(2)} & \Lambda_3^{(3)} \end{bmatrix}\left\{\begin{matrix} \hat{v}_1 \\ \hat{v}_2 \\ \hat{v}_3 \end{matrix}\right\} = \left\{\begin{matrix} v_1 \\ v_2 \\ v_3 \end{matrix}\right\}. \tag{1.38}$$

Remark 2 The collective behavior of the group of *noninteracting* particles, without any particle-to-particle interaction, with zero initial velocities, and no electromagnetic field yields

$$\sum_{i=1}^{N} m_i \dot{\boldsymbol{v}}_i = \sum_{i=1}^{N} q_i \boldsymbol{v}_i \times \boldsymbol{B}^{ext} = \mathbf{0}, \tag{1.39}$$

and the linear momentum of the group and each individual particle remains zero, i.e., nothing moves. This also implies that the angular momentum about any point remains zero (no rotation). However, if the particles all had a uniform initial velocity, the system will accelerate, provided that the overall sum of the charges is not zero, as indicated by

$$\sum_{i=1}^{N} m_i \dot{\boldsymbol{v}}_i = \sum_{i=1}^{N} q_i \boldsymbol{v}_i \times \boldsymbol{B}^{ext} = (\sum_{i=1}^{N} q_i) \boldsymbol{v}_o \times \boldsymbol{B}^{ext}. \tag{1.40}$$

The corresponding angular momentum about the center of mass is

$$\begin{aligned} \sum_{i=1}^{N} \boldsymbol{r}_{cm\to i} \times m_i \dot{\boldsymbol{v}}_i &= \sum_{i=1}^{N} \boldsymbol{r}_{cm\to i} \times q_i \boldsymbol{v}_i \times \boldsymbol{B}^{ext} \\ &= (\sum_{i=1}^{N} \boldsymbol{r}_{cm\to i} q_i) \times \boldsymbol{v}_o \times \boldsymbol{B}^{ext}. \end{aligned} \tag{1.41}$$

Thus, if either

$$\sum_{i=1}^{N} \boldsymbol{r}_{cm\to i} q_i = \mathbf{0} \tag{1.42}$$

or

$$\boldsymbol{v}_o \times \boldsymbol{B}^{ext} = \mathbf{0}, \tag{1.43}$$

or if $\sum_{i=1}^{N} \boldsymbol{r}_{cm\to i} q_i$ is parallel to $\boldsymbol{v}_o \times \boldsymbol{B}^{ext}$, then there will be no rotation of the collection of particles. Clearly, when the particles are uniformly charged, $q_i = q_o$, then

$$\sum_{i=1}^{N} \boldsymbol{r}_{cm\to i} q_i = q_o \sum_{i=1}^{N} \boldsymbol{r}_{cm\to i} = \mathbf{0} \tag{1.44}$$

since the relative position vectors are measured with respect to the center of mass. One can visualize this type of motion as having each particle trace out a circular translational motion, similar to a crank mechanism connecting wheels of a train (Fig. 1.2). No overall rotation of the system occurs, the particles all trace out identical circular paths. Only in the case when the initial velocity of the particles is nonuniform,

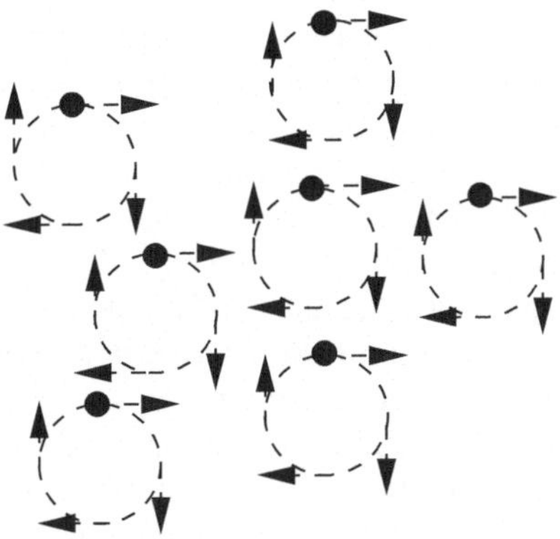

Fig. 1.2 Simultaneous, synchronous, circular paths by each particle leading to no relative (particle-to-particle) and no overall system rotation

or when the particles exhibit particle-to-particle interaction, generating nonuniform velocities, will there be magnetically-induced rotation.

Remark 3 Thus far, we have not considered particle-to-particle interaction. Specifically, next, we will consider a decomposition of the forces governing the dynamics of the particles into an independent "external" part and a particle-to-particle "internal" (near-field) part

$$\underbrace{q_i(\boldsymbol{E}^{ext} + \boldsymbol{v}_i \times \boldsymbol{B}^{ext}) + \sum_{j \neq i}^{N} \boldsymbol{\Psi}_{ij}}_{\text{total}} = \underbrace{q_i(\boldsymbol{E}^{ext} + \boldsymbol{v}_i \times \boldsymbol{B}^{ext})}_{\text{external}} + \underbrace{\sum_{j \neq i}^{N} \boldsymbol{\Psi}_{ij}}_{\text{internal}}. \tag{1.45}$$

1.5 Interparticle Near-Field Interaction

In order to motivate the near-field interaction that is often present between particulates (small-scale particles), we recall that the force between two electrically charged particles is given by (Coulomb's law)

$$\boldsymbol{\Psi}^e_{ij} = \frac{q_i q_j}{4\pi\epsilon||\boldsymbol{r}_i - \boldsymbol{r}_j||^2}\boldsymbol{n}_{ij}, \tag{1.46}$$

where $\boldsymbol{\Psi}^e_{ij}$ is the force acting between the particles, q_i is the charge of particle i, q_j is the charge of particle j, ϵ is the permittivity and $\boldsymbol{n}_{ij}$ is the normal direction determined by the difference in the position vectors of the particles' centers, defined by

$$\boldsymbol{n}_{ij} \stackrel{\text{def}}{=} -\frac{\boldsymbol{r}_i - \boldsymbol{r}_j}{||\boldsymbol{r}_i - \boldsymbol{r}_j||} = \frac{\boldsymbol{r}_j - \boldsymbol{r}_i}{||\boldsymbol{r}_i - \boldsymbol{r}_j||}, \tag{1.47}$$

where $||\boldsymbol{r}_i - \boldsymbol{r}_j||$ is the separation distance between particles i (located at $\boldsymbol{r}_i$) and j (located at $\boldsymbol{r}_j$) and $||\cdot||$ represents the Euclidean norm in R^3. Usually, one writes $\epsilon = \epsilon_o \epsilon_r$ where $\epsilon_o = 8.854 \times 10^{-12}$ farads/meter is the free space permittivity and ϵ_r is the relative permittivity or "dielectric" constant. For point charges of like sign, the Coulombic force is one of repulsion, while for opposite charges, the force is attractive. While, in theory, Maxwell's equations can be applied at the scale of each component within a collection of (charged) particulate ions, the system of equations would become so massive that its solution would be out of reach of virtually all computers available within the near future. For this reason, empirically-generated interaction laws for complex particulate ion-ion interaction, resulting in effective interaction that have attractive and repulsive components, are employed. The electric field induced on a particle i by a particle j, is a result of interaction between complex aggregates of positive and negative charges together. A simple form that captures the essential near-field features is

$$\boldsymbol{\Psi}_i^{nf} = \sum_{j \neq i}^{N_p} \left(\underbrace{\alpha_{1ij} ||\boldsymbol{r}_i - \boldsymbol{r}_j||^{-\beta_1}}_{\text{attraction}} - \underbrace{\alpha_{2ij} ||\boldsymbol{r}_i - \boldsymbol{r}_j||^{-\beta_2}}_{\text{repulsion}} \right) \underbrace{\boldsymbol{n}_{ij}}_{\text{unit vector}} , \qquad (1.48)$$

where the α's and β's are empirical material parameters.

Remark 1 The various representations (decompositions) of the coefficients that appear in Eq. 1.48 are with $c_i = \pm 1$ (a positive/negative identifier)

- mass-based (m = mass) : $\alpha_{ij} = \bar{\alpha}_{ij} m_i m_j c_i c_j$,
- surface area-based (a = surface area) : $\alpha_{ij} = \bar{\alpha}_{ij} a_i a_j c_i c_j$,
- volume-based (V = volume) : $\alpha_{ij} = \bar{\alpha}_{ij} V_i V_j c_i c_j$ and
- charge-based: $\alpha_{ij} = \bar{\alpha}_{ij} q_i q_j c_i c_j$,

where the $\bar{\alpha}_{ij}$ are empirical material parameters. When a particle is relatively large, over essentially a millimeter, near-fields are quite small, in comparison with other forces. However, below one mm, a combination of the various near-field effects, positive and negative charge distributions, etc., can lead to a composite, or "effective" near-field, composed of an attractive and repulsive force (Eq. 1.48).

Remark 2 We will utilize the decomposition of the electromagnetic forces generated into a (interparticle) near-field interaction and the external electromagnetic field

$$\boldsymbol{\Psi}_i = \underbrace{\sum_{j \neq i}^{N} \boldsymbol{\Psi}_{ij}^{nf}}_{\boldsymbol{\Psi}_i^{nf}} + \boldsymbol{\Psi}_i^{env} = \boldsymbol{\Psi}_i^{nf} + \underbrace{q_i(\boldsymbol{E}^{ext} + \boldsymbol{v}_i \times \boldsymbol{B}^{ext})}_{\boldsymbol{\Psi}_i^{env}}, \qquad (1.49)$$

where $\sum_{j \neq i}^{N} \boldsymbol{\Psi}_{ij}^{nf}$ represents the interaction between particle i and all other particles $j = 1, 2...N$ $(j \neq i)$, $\boldsymbol{\Psi}_i^{env}$ represents external ("environment") forces from the surrounding environment, for example, comprised of $\boldsymbol{E}^{ext}$ and $\boldsymbol{B}^{ext}$, which are externally-controlled fields that are independent of the response of the system. $\boldsymbol{E}^{ext}$ and $\boldsymbol{B}^{ext}$ can be considered as static (or extremely slowly-varying), and thus mutually uncoupled and independently controllable. The self-induced magnetic fields developed between particles is insignificant for the velocity ranges of interest here (well below the speed of light). Other forces, arising from contact with friction, will be introduced shortly.

Remark 3 The specific structure of the near-field interaction law chosen was only one of many possibilities to model near-field behavior. There are vast numbers of empirical representations, for example, found in the field of "Molecular Dynamics" (MD), which typically refers to mathematical models of systems of atoms or molecules where each atom (or molecule) is represented by a material point in R^3 and is treated as a point mass. The overall motion of such mass-point systems is dictated by Newtonian mechanics. For an extensive survey of MD-type interaction forces, which includes comparisons of the theoretical and computational properties of a variety of interaction laws, we refer the reader to Frenklach and Carmer [12]. In the usual MD approach (see Haile [17], for example), the motion of individual atoms is described by Newton's second law with the forces computed from differentiating a prescribed potential energy function, with applications to solids, liquids, and gases, as well as biological systems [18, 45, 46]. The interaction functions usually take the form of the familiar Mie, Lennard–Jones, and Morse potentials [40], however three-body terms can be introduced directly into the interaction functions [47] or, alternatively, "local" modifications can be made to two-body representations [51].

Remark 4 The study of uncharged "granular" or "particulate" media, in the absence of electromagnetic effects, is wide ranging. Classical examples include the study of natural materials, such as, sand and gravel, associated with coastal erosion, landslides and avalanches. For reviews, see Pöschel and Schwager [44], Duran [11], Jaeger and Nagel [25, 26], Nagel [42], Liu et al. [39], Liu and Nagel [38], Jaeger and Nagel [27], Jaeger et al. [29–31], Jaeger and Nagel [28], the extensive works of Hutter and collaborators: Tai et al. [48–50], Gray et al. [15], Wieland et al. [53], Berezin et al. [5], Gray and Hutter [14], Gray [13], Hutter [20], Hutter et al. [23], Hutter and Rajagopal [21], Koch et al. [37], Greve and Hutter [16] and Hutter et al. [22]; the works of Behringer and collaborators: Behringer [1], Behringer and Baxter [2], Behringer and Miller [3] and Behringer et al. [4]; the works of Jenkins and collaborators: Jenkins and Strack [34], Jenkins and La Ragione [33], Jenkins and Koenders [32], Jenkins et al. [35] and the works of Torquato and collaborators: Torquato [52], Kansaal et al. [36] and Donev et al. [6–10]. In the manufacturing of particulate composite materials, small-scale particles which are transported and introduced into a molten matrix play a central role. For example, see Hashin [19], Mura [41], Nemat-Nasser and Hori [43],Torquato [52] and Zohdi and Wriggers [54].

References

1. Behringer, R. P. (1993). The dynamics of flowing sand. *Nonlinear Science Today, 3*, 1.
2. Behringer, R. P., & Baxter, G. W. (1993). Pattern formation, complexity and time-dependence in granular flows. In A. Mehta (Ed.), *Granular matter–an interdisciplinary approach* (pp. 85–119). New York: Springer.
3. Behringer, R. P., & Miller, B. J. (1997). Stress fluctuations for sheared 3D granular materials. In Behringer, R., Jenkins, J., (Eds.), *Proceedings, powders and grains 97* pp. 333–336). Balkema.
4. Behringer, R. P., Howell, D., & Veje, C. (1999). Fluctuations in granular flows. *Chaos, 9*, 559–572.
5. Berezin, Y. A., Hutter, K., & Spodareva, L. A. (1998). Stability properties of shallow granular flows. International Journal of Nonlinear Mechanics, 33(4), 647–658.19.
6. Donev, A., Cisse, I., Sachs, D., Variano, E. A., Stillinger, F., Connelly, R., et al. (2004). Improving the density of jammed disordered packings using ellipsoids. *Science, 303*, 990–993.
7. Donev, A., Stillinger, F. H., Chaikin, P. M., & Torquato, S. (2004). Unusually dense crystal ellipsoid packings. *Physics Review Letter, 92*, 255506.
8. Donev, A., Torquato, S., & Stillinger, F. (2005). Neighbor list collision-driven molecular dynamics simulation for nonspherical hard particles-I. *Algorithmic details. Journal of Computer Physics, 202*, 737.
9. Donev, A., Torquato, S., & Stillinger, F. (2005). Neighbor list collision-driven molecular dynamics simulation for nonspherical hard particles-II. Application to ellipses and ellipsoids. Journal of Computer. *Physics, 202*, 765.
10. Donev, A., Torquato, S., & Stillinger, F. H. (2005). Pair correlation function characteristics of nearly jammed disordered and ordered hard-sphere packings. *Physics Review E, 71*, 011105.
11. Duran, J. (1997). *Sands, powders and grains*. An introduction to the physics of granular matter. New York: Springer.
12. Frenklach, M., & Carmer, C. S. (1999). Molecular dynamics using combined quantum and empirical forces: Application to surface reactions. *Advances in Classical Trajectory Methods, 4*, 27–63.
13. Gray, J. M. N. T. (2001). Granular flow in partially filled slowly rotating drums. *Journal of Fluid Mechanics, 441*, 1–29.
14. Gray, J. M. N. T., & Hutter, K. (1997). *Pattern formation in granular avalanches. CMT, 9*, 341–345.
15. Gray, J. M. N. T., Wieland, M., & Hutter, K. (1999). Gravity-driven free surface flow of granular avalanches over complex basal topography. *Proceedings of the Royal Society of London, A, 455*, 1841–1874.
16. Greve, R., & Hutter, K. (1993). Motion of a granular avalanche in a convex and concave curved chute: Experiments and theoretical predictions. *Philos Trans R Soc London A, 342*, 573–600.
17. Haile, J. M. (1992). *Molecular dynamics simulations: Elementary methods*. New York: Wiley.
18. Hase, W. L. (1999). Molecular dynamics of clusters, surfaces, liquids and interfaces. In W. L. Hase (Ed.), *Advances in classical trajectory methods* (Vol. 4). Stamford: JAI Press.
19. Hashin, Z. (1983). Analysis of composite materials: A survey. *ASME Journal of Applied Mechanics, 50*, 481–505.
20. Hutter, K. (1996). Avalanche dynamics. In V. P. Singh (Ed.), *Hydrology of disasters* (pp. 317–394). Dordrecht: Kluwer Academic Publishers.
21. Hutter, K., & Rajagopal, K. R. (1994). *On flows of granular materials. CMT, 6*, 81–139.
22. Hutter, K., Siegel, M., Savage, S. B., & Nohguchi, Y. (1993). Two-dimensional spreading of a granular avalanche down an inclined plane. *Part I: Theory. Acta Mechanica, 100*, 37–68.
23. Hutter, K., Koch, T., Plüss, C., & Savage, S. B. (1995). The dynamics of avalanches of granular materials from initiation to runout. *Part II. Experiments. Acta Mechanica, 109*, 127–165.
24. Jackson, J. D. (1998). *Classical electrodynamics* (3rd ed.). New York: Wiley.

25. Jaeger, H. M., & Nagel, S. R. (1992). La Physique de l'Etat Granulaire. *La Recherche, 249*, 1380.
26. Jaeger, H. M., & Nagel, S. R. (1992). Physics of the granular state. *Science, 255*, 1523.
27. Jaeger, H. M., & Nagel, S. R. (1993). La Fisica del Estado granular. *Mundo Cientifico, 132*, 108.
28. Jaeger, H. M., & Nagel, S. R. (1997). Dynamics of granular material. *American Scientist, 85*, 540.
29. Jaeger, H. M., Knight, J. B., Liu, C. H., & Nagel, S. R. (1994). What is shaking in the sand box? *Materials Research Society Bulletin, 19*, 25.
30. Jaeger, H. M., Nagel, S. R., & Behringer, R. P. (1996). The physics of granular materials. *Physics Today, 4*, 32.
31. Jaeger, H. M., Nagel, S. R., & Behringer, R. P. (1996). Granular solids, liquids and gases. *Reviews of Modern Physics, 68*, 1259.
32. Jenkins, J. T., & Koenders, M. A. (2004). The incremental response of random aggregates of identical round particles. *European Physical Journal E - Soft Matter, 13*, 113–123.
33. Jenkins, J. T., & La Ragione, L. (1999). Particle spin in anisotropic granular materials. *International Journal of Solids and Structures, 38*, 1063–1069.
34. Jenkins, J. T., & Strack, O. D. L. (1993). Mean-field inelastic behavior of random arrays of identical spheres. *JoMMS, 16*, 25–33.
35. Jenkins, J. T., Johnson, D., La Ragione, L., & Makse, H. (2005). Fluctuations and the effective moduli of an isotropic, random aggregate of identical, frictionless spheres. Journal of the Mechanics and Physics of Solids, 197–225.
36. Kansaal, A., Torquato, S., & Stillinger, F. (2002). Diversity of order and densities in jammed hard-particle packings. *Physics Review E, 66*, 041109.
37. Koch, T., Greve, R., & Hutter, K. (1994). Unconfined flow of granular avalanches along a partly curved surface. II. Experiments and numerical computations. *Proceedings of the Royal Society London, A, 445*, 415–435.
38. Liu, C. H., & Nagel, S. R. (1993). Sound in a granular material: Disorder and nonlinearity. *Physics Review B, 48*, 15646.
39. Liu, C. H., Jaeger, H. M., & Nagel, S. R. (1991). Finite size effects in a Sandpile. *Physics Review A, 43*, 7091.
40. Moelwyn-Hughes, E. A. (1961). *Physical Chemistry*. New York: Pergamon.
41. Mura, T. (1993). Micromechanics of defects in solids (2nd ed.). Kluwer Academic Publishers.
42. Nagel, S. R. (1992). Instabilities in a Sandpile. *Reviews of Modern Physics, 64*, 321.
43. Nemat-Nasser, S., & Hori, M. (1999). *Micromechanics: overall properties of heterogeneous solids* (2nd ed.). Amsterdam: Elsevier.
44. Pöschel, T., & Schwager, T. (2004). *Computational granular dynamics*. New York: Springer.
45. Rapaport, D. C. (1995). *The art of molecular dynamics simulation*. Cambridge: Cambridge University Press.
46. Schlick, T. (2000). *Molecular modeling and simulation: An interdisciplinary guide*. New York: Springer.
47. Stillinger, F. H., & Weber, T. A. (1985). Computer simulation of local order in condensed phases of silicon. *Physical Review B, 31*, 5262–5271.
48. Tai, Y.-C., Gray, J. M. N. T., Hutter, K., & Noelle, S. (2001). Flow of dense avalanches past obstructions. *Annals of Glaciology, 32*, 281–284.
49. Tai, Y.-C., Noelle, S., Gray, J. M. N. T., & Hutter, K. (2001). An accurate shock-capturing finite-difference method to solve the Savage-Hutter equations in avalanche dynamics. *Annals of Glaciology, 32*, 263–267.
50. Tai, Y.-C., Noelle, S., Gray, J. M. N. T., & Hutter, K. (2002). Shock capturing and front tracking methods for granular avalanches. *Journal of Computer Physics, 175*, 269–301.
51. Tersoff, J. (1988). Empirical interatomic potential for carbon, with applications to amorphous carbon. *Physical Review Letter, 61*, 2879–2882.
52. Torquato, S. (2002). *Random Heterogeneous Materials: Microstructure and Macroscopic Properties*. New York: Springer.

53. Wieland, M., Gray, J. M. N. T., & Hutter, K. (1999). Channelized free-surface flow of cohesionless granular avalanches in a chute with shallow lateral curvature. *Journal of Fluid Mechanics, 392*, 73–100.
54. Zohdi, T. I., & Wriggers, P. (2008). *Introduction to computational micromechanics, second Reprinting*. Springer.

Chapter 2
Dynamics of Rigid Clusters of Charged Particles

As the next level of complexity beyond the dynamics of a single particle, we consider rigid clusters of such particles. In this chapter, we consider the cluster to already be formed, with particles rigidly bound together by either mechanical, chemical or electromagnetic bonds. Investigation of the evolution of such clusters from loose, free-flowing, particulates will be discussed later in the monograph. Of particular interest is to compare and contrast the differences in the dynamics of a cluster of charged particles and that of a (hypothetical) single charged particle (with the same overall charge) whose motion is governed by

$$m\dot{\boldsymbol{v}} = q(\boldsymbol{E}^{ext} + \boldsymbol{v} \times \boldsymbol{B}^{ext}), \tag{2.1}$$

where m is the mass of the particle, $\boldsymbol{v}$ is the particle velocity, $\boldsymbol{E}^{ext} = (E_1^{ext}, E_2^{ext}, E_3^{ext})$ is the external electric field and $\boldsymbol{B}^{ext} = (B_1^{ext}, B_2^{ext}, B_3^{ext})$ is the external magnetic field.

Remark Such clusters possess unique dynamics that are important in order to understand and fully control relevant industrial processes (see, for example, Luo and Dornfeld [14–17], Arbelaez et al. [1, 2], Ciampini et al. [3, 4], Gomes-Ferreira et al. [10], Ghobeity et al. [7, 8] and Zohdi [27–45]).[1]

2.1 Dynamics of Charged Clusters

Following an approach found in Zohdi [44], consider a collection of rigidly-bonded particles, $i = 1, 2, ..., N_c$, in a cluster. The individual particle dynamics are described by (which leads to a coupled system)

[1] For a review of the effects of clusters on the macroscale material properties of solids that contain them, see Torquato [25], as well as Ghosh et al. [9] for domain partitioning methods that are capable of handling materials with general nonuniform microstructure. For a review of novel approaches of multiscale methods that bridge scales with applications to nanotechnology, see Fish [5].

T. I. Zohdi, *Dynamics of Charged Particulate Systems*, SpringerBriefs in Applied Sciences and Technology, DOI: 10.1007/978-3-642-28519-6_2,

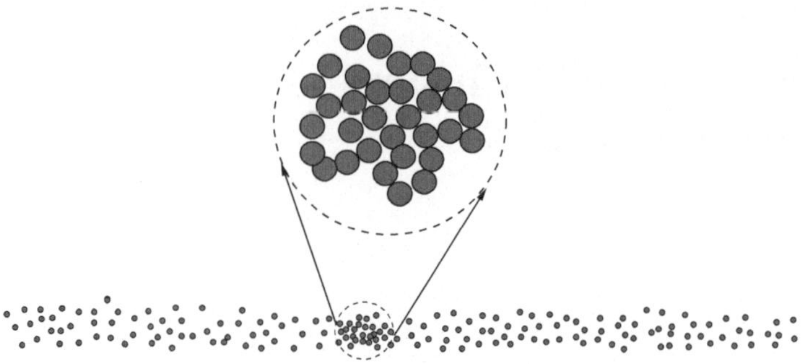

Fig. 2.1 Agglomeration of material (formation of a cluster) within a flow of particulates (Zohdi [44])

$$m_i\ddot{\boldsymbol{r}}_i = \underbrace{\boldsymbol{\Psi}_i^{tot}}_{\text{total forces}} = \underbrace{\boldsymbol{\Psi}_i^{int}}_{\text{internal}} + \underbrace{\boldsymbol{\Psi}_i^{ext}}_{\text{external}} = \boldsymbol{\Psi}_i^{int} + \underbrace{q_i(\boldsymbol{E}^{ext} + \boldsymbol{v}_i \times \boldsymbol{B}^{ext})}_{\boldsymbol{\Psi}_i^{ext}}, \tag{2.2}$$

where $\boldsymbol{r}_i$ is the position vector of the ith particle, m_i is the mass of a single particle and $\boldsymbol{\Psi}_i^{tot}$ is the sum of the forces acting on the ith particle, due to other particles in the system ("internal" particle-to-particle near-fields, bonding forces, etc., $\boldsymbol{\Psi}_i^{int}$) and due to the external electric and magnetic fields ($\boldsymbol{\Psi}_i^{ext}$) (Fig. 2.1).

Remarks Although the exact nature of particle-to-particle interaction is not important in the present (overall rigid motion) analysis (it will be later) since the corresponding forces are internal to the system, in passing, we mention that there are a variety of possible interparticle representations for loose, free-flowing, charged particles. We again refer the reader to Frenklach and Carmer [6] as well as to Haile [11], Hase [12], Schlick [22], Rapaport [19], Torquato [26], Rechtsman et al. [20, 21] and Zohdi [27–45] for overviews of the various representations for particle interaction, for example, those based on the familiar Mie, Lennard–Jones, and Morse potentials (see Moelwyn-Hughes [18] for reviews). Also, three-body terms can be introduced directly into the interparticle interaction (Stillinger [23]) or via term-wise modifications to the two-body representations (Tersoff [24]).

2.1.1 Group Dynamics of a Rigidly Bound Collection of Particles

When we consider a collection of particles that are bound together as a rigid body, the exact nature of the internal particle-to-particle interaction is irrelevant to the overall system dynamics since the internal forces in the system are equal in magnitude and opposite in direction, leading to

$$\sum_{i=1}^{N_c}(\boldsymbol{\Psi}_i^{ext} + \boldsymbol{\Psi}_i^{int}) = \sum_{i=1}^{N_c}\boldsymbol{\Psi}_i^{ext} + \underbrace{\sum_{i=1}^{N_c}\boldsymbol{\Psi}_i^{int}}_{=\boldsymbol{0}} = \sum_{i=1}^{N_c}\boldsymbol{\Psi}_i^{ext} \stackrel{\text{def}}{=} \boldsymbol{\Psi}^{EXT}, \tag{2.3}$$

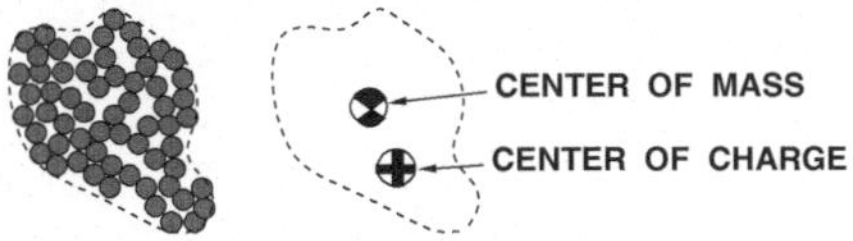

Fig. 2.2 A collection of charged particles that are rigidly bonded together. The nature of the particle-to-particle bonding (mechanical, electronic, chemical, etc.) is irrelevant in the present analysis. The mass center and "charge" center will generally not coincide. This difference will lead to a variation in the dynamics of the center-of-mass of a cluster relative to a single charged particle of equal mass and charge, strongly influenced by the equation of overall balance of angular momentum (Zohdi [44])

where $\boldsymbol{\Psi}^{EXT}$ is the overall external force acting on the cluster and N_c are the number of particles in the cluster (Fig. 2.2). The position vector of the center of mass of the system is given by

$$\boldsymbol{r}_{cm} \stackrel{\text{def}}{=} \frac{\sum_{i=1}^{N_c} m_i \boldsymbol{r}_i}{\sum_{i=1}^{N_c} m_i} = \frac{1}{\mathcal{M}} \sum_{i=1}^{N_c} m_i \boldsymbol{r}_i, \tag{2.4}$$

where $\mathcal{M}$ is the total system mass. A decomposition of the position vector for particle i, of the form $\boldsymbol{r}_i = \boldsymbol{r}_{cm} + \boldsymbol{r}_{cm\to i}$, allows the linear momentum of the system of particles ($\boldsymbol{G}$) to be written as

$$\sum_{i=1}^{N_c} \underbrace{m_i \dot{\boldsymbol{r}}_i}_{\boldsymbol{G}_i} = \sum_{i=1}^{N_c} m_i (\dot{\boldsymbol{r}}_{cm} + \dot{\boldsymbol{r}}_{cm\to i}) = \sum_{i=1}^{N_c} m_i \dot{\boldsymbol{r}}_{cm} = \dot{\boldsymbol{r}}_{cm} \sum_{i=1}^{N_c} m_i = \mathcal{M} \dot{\boldsymbol{r}}_{cm} \stackrel{\text{def}}{=} \boldsymbol{G}_{cm} \tag{2.5}$$

since $\sum_{i=1}^{N_c} m_i \dot{\boldsymbol{r}}_{cm\to i} = \boldsymbol{0}$. Furthermore, $\dot{\boldsymbol{G}}_{cm} = \mathcal{M} \ddot{\boldsymbol{r}}_{cm}$, and thus

$$\dot{\boldsymbol{G}}_{cm} = \mathcal{M} \ddot{\boldsymbol{r}}_{cm} = \sum_{i=1}^{N_c} \psi_i^{ext} = \sum_{i=1}^{N_c} q_i (\boldsymbol{E}^{ext} + \boldsymbol{v}_i \times \boldsymbol{B}^{ext}) \stackrel{\text{def}}{=} \boldsymbol{\Psi}^{EXT}. \tag{2.6}$$

The angular momentum relative to the center of mass can be written as (utilizing $\dot{\boldsymbol{r}}_i = \boldsymbol{v}_i = \boldsymbol{v}_{cm} + \boldsymbol{v}_{cm\to i}$)

$$\sum_{i=1}^{N_c} = \boldsymbol{H}_{cm\to i} = \sum_{i=1}^{N_c} (\boldsymbol{r}_{cm\to i} \times m_i \boldsymbol{v}_{cm\to i}) = \sum_{i=1}^{N_c} (\boldsymbol{r}_{cm\to i} \times m_i (\boldsymbol{v}_i - \boldsymbol{v}_{cm})) \tag{2.7}$$

$$= \sum_{i=1}^{N_c} (m_i \boldsymbol{r}_{cm\to i} \times \boldsymbol{v}_i) - \left(\underbrace{\sum_{i=1}^{N_c} m_i \boldsymbol{r}_{cm\to i}}_{=0} \right) \times \boldsymbol{v}_{cm} = \boldsymbol{H}_{cm} \tag{2.8}$$

for a rigid body. Since $\boldsymbol{v}_{cm\to i} = \boldsymbol{\omega} \times \boldsymbol{r}_{cm\to i}$,

$$\boldsymbol{H}_{cm} = \sum_{i=1}^{N_c} \boldsymbol{H}_{cm\to i} = \sum_{i=1}^{N_c} m_i(\boldsymbol{r}_{cm\to i} \times \boldsymbol{v}_{cm\to i}) = \sum_{i=1}^{N_c} m_i(\boldsymbol{r}_{cm\to i} \times (\boldsymbol{\omega} \times \boldsymbol{r}_{cm\to i})). \tag{2.9}$$

Decomposing the relative position vector into its components

$$\boldsymbol{r}_{cm\to i} = \boldsymbol{r}_i - \boldsymbol{r}_{cm} = \hat{x}_{i1}\boldsymbol{e}_1 + \hat{x}_{i2}\boldsymbol{e}_2 + \hat{x}_{i3}\boldsymbol{e}_3, \tag{2.10}$$

where $\hat{x}_{i1}$, $\hat{x}_{i2}$ and $\hat{x}_{i3}$ are the coordinates of the mass points measured *relative to the center of mass*, and expanding the angular momentum expression, yields

$$H_1 = \omega_1 \sum_{i=1}^{N_c} (\hat{x}_{i2}^2 + \hat{x}_{i3}^2)\, m_i - \omega_2 \sum_{i=1}^{N_c} \hat{x}_{i1}\hat{x}_{i2}\, m_i - \omega_3 \sum_{i=1}^{N_c} \hat{x}_{i1}\hat{x}_{i3}\, m_i \tag{2.11}$$

$$H_2 = -\omega_1 \sum_{i=1}^{N_c} \hat{x}_{i1}\hat{x}_{i2}\, m_i + \omega_2 \sum_{i=1}^{N_c} (\hat{x}_{i1}^2 + \hat{x}_{i3}^2)\, m_i - \omega_3 \sum_{i=1}^{N_c} \hat{x}_{i2}\hat{x}_{i3}\, m_i \tag{2.12}$$

and

$$H_3 = -\omega_1 \sum_{i=1}^{N_c} \hat{x}_{i1}\hat{x}_{i3}\, m_i - \omega_2 \sum_{i=1}^{N_c} \hat{x}_{i2}\hat{x}_{i3}\, m_i + \omega_3 \sum_{i=1}^{N_c} (\hat{x}_{i1}^2 + \hat{x}_{i2}^2)\, m_i, \tag{2.13}$$

which can be concisely written as

$$\boldsymbol{H}_{cm} = \overline{\boldsymbol{\mathcal{I}}} \cdot \boldsymbol{\omega}, \tag{2.14}$$

where we define the moments of inertia with respect to the center of mass

$$\overline{\mathcal{I}}_{11} = \sum_{i=1}^{N_c} (\hat{x}_{i2}^2 + \hat{x}_{i3}^2)\, m_i, \quad \overline{\mathcal{I}}_{22} = \sum_{i=1}^{N_c} (\hat{x}_{i1}^2 + \hat{x}_{i3}^2)\, m_i, \quad \overline{\mathcal{I}}_{33} = \sum_{i=1}^{N_c} (\hat{x}_{i1}^2 + \hat{x}_{i2}^2)\, m_i, \tag{2.15}$$

$$\overline{\mathcal{I}}_{12} = \overline{\mathcal{I}}_{21} = -\sum_{i=1}^{N_c} \hat{x}_{i1}\hat{x}_{i2}\, m_i, \; \overline{\mathcal{I}}_{23} = \overline{\mathcal{I}}_{32} = -\sum_{i=1}^{N_c} \hat{x}_{i2}\hat{x}_{i3}\, m_i,$$

$$\overline{\mathcal{I}}_{13} = \overline{\mathcal{I}}_{31} = -\sum_{i=1}^{N_c} \hat{x}_{i1}\hat{x}_{i3}\, m_i, \tag{2.16}$$

or explicitly

$$\overline{\boldsymbol{\mathcal{I}}} = \begin{bmatrix} \overline{\mathcal{I}}_{11} & \overline{\mathcal{I}}_{12} & \overline{\mathcal{I}}_{13} \\ \overline{\mathcal{I}}_{21} & \overline{\mathcal{I}}_{22} & \overline{\mathcal{I}}_{23} \\ \overline{\mathcal{I}}_{31} & \overline{\mathcal{I}}_{32} & \overline{\mathcal{I}}_{33} \end{bmatrix}. \tag{2.17}$$

The particles' own inertia contribution about their respective mass-centers to the overall moment of inertia of the agglomerated body can be described by the Huygens–Steiner (generalized "parallel axis" theorem) formula ($p, s = 1, 2, 3$)

$$\bar{\mathcal{I}}_{ps} = \sum_{i=1}^{N_c} \left(\bar{\mathcal{I}}^i_{ps} + m_i (||\boldsymbol{r}_i - \boldsymbol{r}_{cm}||^2 \delta_{ps} - \hat{x}_{ip}\hat{x}_{is}) \right). \tag{2.18}$$

For a spherical particle, $\bar{\mathcal{I}}^i_{pp} = \frac{2}{5} m_i R_i^2$, and for $p \neq s$, $\bar{\mathcal{I}}^i_{ps} = 0$ (no products of inertia), R_i being the particle radius.[2] Finally, for the derivative of the angular momentum, utilizing $\ddot{\boldsymbol{r}}_i = \boldsymbol{a}_i = \boldsymbol{a}_{cm} + \boldsymbol{a}_{cm\to i}$, we obtain

$$\dot{\boldsymbol{H}}^{rel}_{cm} = \sum_{i=1}^{N_c} (\boldsymbol{r}_{cm\to i} \times m_i \boldsymbol{a}_{cm\to i}) = \sum_{i=1}^{N_c} (\boldsymbol{r}_{cm\to i} \times m_i (\boldsymbol{a}_i - \boldsymbol{a}_{cm})) \tag{2.19}$$

$$= \sum_{i=1}^{N_c} (m_i \boldsymbol{r}_{cm\to i} \times \boldsymbol{a}_i) - \underbrace{(\sum_{i=1}^{N_c} m_i \boldsymbol{r}_{cm\to i})}_{=0} \times \boldsymbol{a}_{cm} = \dot{\boldsymbol{H}}_{cm}, \tag{2.20}$$

and consequently

$$\dot{\boldsymbol{H}}_{cm} = \frac{d(\bar{\mathcal{I}} \cdot \boldsymbol{\omega})}{dt} = \sum_{i=1}^{N_c} \boldsymbol{r}_{cm\to i} \times \boldsymbol{\Psi}^{ext}_i = \sum_{i=1}^{N_c} \boldsymbol{r}_{cm\to i} \times q_i (\boldsymbol{E}^{ext} + \boldsymbol{v}_i \times \boldsymbol{B}^{ext}) \stackrel{\text{def}}{=} \boldsymbol{M}^{EXT}_{cm}, \tag{2.21}$$

where $\boldsymbol{M}^{EXT}_{cm}$ is the total external moment about the center of mass.

2.2 Decomposition of the Electromagnetic Contributions

Consider a rigid cluster of charged particles with angular velocity $\boldsymbol{\omega}$ and center of mass velocity $\boldsymbol{v}_{cm}$.

2.2.1 The Overall Forces and Moments

The velocity of any point on the body can be represented by

$$\boldsymbol{v}_i = \boldsymbol{v}_{cm} + \boldsymbol{\omega} \times \boldsymbol{r}_{cm\to i}, \tag{2.22}$$

[2] If the particles are sufficiently small, each particle's own moment inertia (about its own center) is insignificant, leading to $\bar{\mathcal{I}}_{ps} = \sum_{i=1}^{N_c} m_i (||\boldsymbol{r}_i - \boldsymbol{r}_{cm}||^2 \delta_{ps} - \hat{x}_{ip}\hat{x}_{is})$.

and the overall external electromagnetic force $\boldsymbol{\Psi}^{EXT} = \sum_{i=1}^{N_c} q_i(\boldsymbol{E}^{ext} + \boldsymbol{v} \times \boldsymbol{B}^{ext})$ can be decomposed into the following parts

$$\begin{aligned}
\boldsymbol{\Psi}^{EXT} = & \underbrace{\sum_{i=1}^{N_c} q_i \boldsymbol{E}^{ext}}_{\text{electrical contribution}} + \underbrace{\sum_{i=1}^{N_c} q_i(\boldsymbol{v}_{cm} \times \boldsymbol{B}^{ext})}_{\text{linear velocity contribution}} + \underbrace{\sum_{i=1}^{N_c} q_i((\boldsymbol{\omega} \times \boldsymbol{r}_{cm\to i}) \times \boldsymbol{B}^{ext})}_{\text{angular velocity contribution}} \\
= & \boldsymbol{E}^{ext}\left(\sum_{i=1}^{N_c} q_i\right) + \boldsymbol{v}_{cm} \times \boldsymbol{B}^{ext}\left(\sum_{i=1}^{N_c} q_i\right) + \boldsymbol{\omega} \times \left(\sum_{i=1}^{N_c} q_i \boldsymbol{r}_{cm\to i}\right) \times \boldsymbol{B}^{ext},
\end{aligned} \tag{2.23}$$

and, similarly, for the total external moment about the center of mass,

$$\begin{aligned}
\boldsymbol{M}_{cm}^{EXT} = & \sum_{i=1}^{N_c} \boldsymbol{r}_{cm\to i} \times \Bigg(\underbrace{q_i \boldsymbol{E}^{ext}}_{\text{electrical contribution}} + \underbrace{q_i(\boldsymbol{v}_{cm} \times \boldsymbol{B}^{ext})}_{\text{linear velocity contribution}} \\
& + \underbrace{q_i((\boldsymbol{\omega} \times \boldsymbol{r}_{cm\to i}) \times \boldsymbol{B}^{ext})}_{\text{angular velocity contribution}} \Bigg) \\
= & \left(\sum_{i=1}^{N_c} q_i \boldsymbol{r}_{cm\to i}\right) \times \boldsymbol{E}^{ext} + \left(\sum_{i=1}^{N_c} q_i \boldsymbol{r}_{cm\to i}\right) \times \boldsymbol{v}_{cm} \times \boldsymbol{B}^{ext} \\
& + \left(\sum_{i=1}^{N_c} q_i \boldsymbol{r}_{cm\to i} \times \boldsymbol{\omega} \times \boldsymbol{r}_{cm\to i}\right) \times \boldsymbol{B}^{ext} \\
= & \boldsymbol{R}_q \times \boldsymbol{E}^{ext} + \boldsymbol{R}_q \times \boldsymbol{v}_{cm} \times \boldsymbol{B}^{ext} + \boldsymbol{H}_q \times \boldsymbol{B}^{ext},
\end{aligned} \tag{2.24}$$

where

- $\boldsymbol{R}_q \overset{\text{def}}{=} \sum_{i=1}^{N_c} q_i \boldsymbol{r}_{cm\to i}$ is defined as the *center of charge relative to the center of mass* and
- $\boldsymbol{H}_q \overset{\text{def}}{=} \sum_{i=1}^{N_c} q_i \frac{\boldsymbol{H}_{cm\to i}}{m_i}$ is defined as the *charged angular momentum per unit mass with respect to the center of mass.*

Thus, the following three quantities play a central role in the cluster behavior:

- The sum of the individual charges ("overall charge"/first moment): $Q \overset{\text{def}}{=} \sum_{i=1}^{N_c} q_i$,
- The sum of the distances between the individual charged particles ($\boldsymbol{r}_i$) and the center of mass of the cluster ($\boldsymbol{r}_{cm}$), weighted by the individual charges ("charged radius"/second-moment): $\boldsymbol{R}_q \overset{\text{def}}{=} \sum_{i=1}^{N_c} q_i(\boldsymbol{r}_i - \boldsymbol{r}_{cm})$,
- The sum of the self-cross-product of the distances between the individual charged particles and the center of mass of the cluster, weighted by the individual charges ($\overline{\boldsymbol{\mathcal{I}}}_q$, "moment of charge"/third-moment): $\boldsymbol{H}_q = \overline{\boldsymbol{\mathcal{I}}}_q \cdot \boldsymbol{\omega} \overset{\text{def}}{=} \sum_{i=1}^{N_c} q_i(\boldsymbol{r}_i - \boldsymbol{r}_{cm}) \times$

$\boldsymbol{\omega} \times (\boldsymbol{r}_i - \boldsymbol{r}_{cm})$, where $\boldsymbol{\omega}$ is the angular velocity of the body and where $\overline{\boldsymbol{\mathcal{I}}}_q$ has components $(p, s = 1, 2, 3)$

$$\bar{\mathcal{I}}_{q,ps} = \sum_{i=1}^{N_c} q_i(||\boldsymbol{r}_i - \boldsymbol{r}_{cm}||^2 \delta_{ps} - \hat{x}_{ip}\hat{x}_{is}). \tag{2.25}$$

2.2.2 Various Charge Distribution Cases

In summary, for a charged cluster, the governing equations may be written as

$$\mathcal{M}\ddot{\boldsymbol{r}}_{cm} = \mathcal{M}\dot{\boldsymbol{v}}_{cm} = \boldsymbol{\Psi}^{EXT} = \underbrace{Q\boldsymbol{E}^{ext}}_{\boldsymbol{\mathcal{T}}_1} + \underbrace{Q\boldsymbol{v}_{cm} \times \boldsymbol{B}^{ext}}_{\boldsymbol{\mathcal{T}}_2} + \underbrace{(\boldsymbol{\omega} \times \boldsymbol{R}_q) \times \boldsymbol{B}^{ext}}_{\boldsymbol{\mathcal{T}}_3}, \tag{2.26}$$

and

$$\dot{\boldsymbol{H}}_{cm} = \frac{d(\overline{\boldsymbol{\mathcal{I}}} \cdot \boldsymbol{\omega})}{dt} = \boldsymbol{M}_{cm}^{EXT} = \underbrace{\boldsymbol{R}_q \times \boldsymbol{E}^{ext}}_{\boldsymbol{\mathcal{T}}_4} + \underbrace{\boldsymbol{R}_q \times \boldsymbol{v}_{cm} \times \boldsymbol{B}^{ext}}_{\boldsymbol{\mathcal{T}}_5} + \underbrace{(\overline{\boldsymbol{\mathcal{I}}}_q \cdot \boldsymbol{\omega}) \times \boldsymbol{B}^{ext}}_{\boldsymbol{\mathcal{T}}_6}. \tag{2.27}$$

One may observe that:

- In the special case when the overall charge (Q) of the cluster is zero (neutral), $\boldsymbol{\mathcal{T}}_1 = \boldsymbol{\mathcal{T}}_2 = 0$,
- In the special case when the overall charged distances are evenly distributed with respect to the mass center, $\boldsymbol{R}_q = \boldsymbol{0} \Rightarrow \boldsymbol{\mathcal{T}}_3 = \boldsymbol{\mathcal{T}}_4 = \boldsymbol{\mathcal{T}}_5 = \boldsymbol{0}$,
- In the special case when the overall charged moment ($\overline{\boldsymbol{\mathcal{I}}}_q$) is zero, $\boldsymbol{\mathcal{T}}_6 = 0$.

Also, one has

$$\begin{aligned} ||\boldsymbol{\Psi}^{EXT}|| &= ||Q(\boldsymbol{E}^{ext} + \boldsymbol{v}_{cm} \times \boldsymbol{B}^{ext}) + (\boldsymbol{\omega} \times \boldsymbol{R}_q) \times \boldsymbol{B}^{ext}|| \\ &\leq |Q|||\boldsymbol{E}^{ext} + \boldsymbol{v}_{cm} \times \boldsymbol{B}^{ext}|| + ||\boldsymbol{R}_q||||\boldsymbol{\omega}||||\boldsymbol{B}^{ext}|| \\ &\leq |Q|||\boldsymbol{E}^{ext}|| + |Q|||\boldsymbol{v}_{cm}||||\boldsymbol{B}^{ext}|| + ||\boldsymbol{R}_q||||\boldsymbol{\omega}||||\boldsymbol{B}^{ext}|| \end{aligned} \tag{2.28}$$

and

$$\begin{aligned} ||\boldsymbol{M}_{cm}^{EXT}|| &= ||\boldsymbol{R}_q \times (\boldsymbol{E}^{ext} + \boldsymbol{v}_{cm} \times \boldsymbol{B}^{ext}) + (\overline{\boldsymbol{\mathcal{I}}}_q \cdot \boldsymbol{\omega}) \times \boldsymbol{B}^{ext}|| \\ &\leq ||\boldsymbol{R}_q||||\boldsymbol{E}^{ext} + \boldsymbol{v}_{cm} \times \boldsymbol{B}^{ext}|| + ||\overline{\boldsymbol{\mathcal{I}}}_q||||\boldsymbol{\omega}||||\boldsymbol{B}^{ext}|| \\ &\leq ||\boldsymbol{R}_q||||\boldsymbol{E}^{ext}|| + ||\boldsymbol{R}_q||||\boldsymbol{v}_{cm}||||\boldsymbol{B}^{ext}|| + ||\overline{\boldsymbol{\mathcal{I}}}_q||||\boldsymbol{\omega}||||\boldsymbol{B}^{ext}||. \end{aligned} \tag{2.29}$$

Thus, both Q and $||\boldsymbol{R}_q||$ must be zero for $||\boldsymbol{\Psi}^{EXT}|| = 0$, while both $||\boldsymbol{R}_q||$ and $||\overline{\mathcal{I}}_q||$ must be zero for $||\boldsymbol{M}_{cm}^{EXT}|| = 0$. Clearly, each of the (zero/nonzero) cases can occur independently of one another.

Remark The dynamics of a general cluster must be treated numerically, particularly when one has a three-dimensional body with a complex charge distribution. This is discussed next.

2.3 Numerical Methods for the Dynamics of a Charged Cluster

We now treat the dynamics of a cluster numerically. We first focus on the translational motion of the center of mass, and then turn to the rotational contribution.

2.3.1 Cluster Translational Contribution

The translational component of the center of mass can be written as

$$\mathcal{M}\ddot{\boldsymbol{r}}_{cm} = \mathcal{M}\dot{\boldsymbol{v}}_{cm} = \boldsymbol{\Psi}^{EXT}. \tag{2.30}$$

A trapezoidal time-stepping rule is used, whereby at some intermediate moment in time $t \le t + \phi\Delta t \le t + \Delta t$ $(0 \le \phi \le 1)$,

$$\dot{\boldsymbol{v}}_{cm}(t + \phi\Delta t) \approx \frac{\boldsymbol{v}_{cm}(t + \Delta t) - \boldsymbol{v}_{cm}(t)}{\Delta t} \tag{2.31}$$

$$= \frac{1}{\mathcal{M}}\boldsymbol{\Psi}^{EXT}(t + \phi\Delta t) \tag{2.32}$$

$$\approx \frac{1}{\mathcal{M}}\left(\phi\boldsymbol{\Psi}^{EXT}(t + \Delta t) + (1 - \phi)\boldsymbol{\Psi}^{EXT}(t)\right), \tag{2.33}$$

leading to

$$\boldsymbol{v}_{cm}(t + \Delta t) = \boldsymbol{v}_{cm}(t) + \frac{\Delta t}{\mathcal{M}}\left(\phi\boldsymbol{\Psi}^{EXT}(t + \Delta t) + (1 - \phi)\boldsymbol{\Psi}^{EXT}(t)\right). \tag{2.34}$$

For the position, we have

$$\begin{aligned}\dot{\boldsymbol{r}}_{cm}(t + \phi\Delta t) &\approx \frac{\boldsymbol{r}_{cm}(t + \Delta t) - \boldsymbol{r}_{cm}(t)}{\Delta t} \approx \boldsymbol{v}_{cm}(t + \phi\Delta t)\\ &\approx (\phi\boldsymbol{v}_{cm}(t + \Delta t) + (1 - \phi)\boldsymbol{v}_{cm}(t)),\end{aligned} \tag{2.35}$$

leading to

$$\boldsymbol{r}_{cm}(t+\Delta t) = \boldsymbol{r}_{cm}(t) + \Delta t\left(\phi \boldsymbol{v}_{cm}(t+\Delta t) + (1-\phi)\boldsymbol{v}_{cm}(t)\right). \tag{2.36}$$

2.3.2 Cluster Rotational Motion

There are two possible approaches to compute the cluster rotations, either using an (1) inertially-fixed frame or (2) a body-fixed frame. We employ an inertially-fixed approach, and implicit time-stepping for the duration of this chapter. This straightforward approach entails, at each (implicit) time step, decomposing an increment of motion into an incremental rigid body translational contribution and an incremental rigid body rotational contribution (rotation about the center of mass). The rotational contribution is determined by solving a set of coupled nonlinear equations governing the angular velocity and the incremental rotation of the body around the axis of rotation (which also changes as a function of time). The equation for the angular momentum can be written as

$$\dot{\boldsymbol{H}}_{cm} = \frac{d(\overline{\boldsymbol{\mathcal{I}}}\cdot\boldsymbol{\omega})}{dt} = \boldsymbol{M}_{cm}^{EXT}. \tag{2.37}$$

Because the body rotates, $\overline{\boldsymbol{\mathcal{I}}}$ is implicitly dependent on $\boldsymbol{\omega}$ (and hence time), which leads to a coupled system of nonlinear ODE's which can be solved with an iterative scheme. Equation 2.37 is discretized by a trapezoidal scheme (as for the translational component)

$$\frac{d(\overline{\boldsymbol{\mathcal{I}}}\cdot\boldsymbol{\omega})}{dt}|_{t+\phi\Delta t} = \frac{(\overline{\boldsymbol{\mathcal{I}}}\cdot\boldsymbol{\omega})|_{t+\Delta t} - (\overline{\boldsymbol{\mathcal{I}}}\cdot\boldsymbol{\omega})|_{t}}{\Delta t}, \tag{2.38}$$

thus leading to

$$(\overline{\boldsymbol{\mathcal{I}}}\cdot\boldsymbol{\omega})|_{t+\Delta t} = (\overline{\boldsymbol{\mathcal{I}}}\cdot\boldsymbol{\omega})|_{t} + \Delta t\boldsymbol{M}_{cm}^{EXT}(t+\phi\Delta t). \tag{2.39}$$

Solving for $\boldsymbol{\omega}(t+\Delta t)$ yields

$$\boldsymbol{\omega}(t+\Delta t) = \left(\overline{\boldsymbol{\mathcal{I}}}(t+\Delta t)\right)^{-1}\cdot\left((\overline{\boldsymbol{\mathcal{I}}}\cdot\boldsymbol{\omega})|_{t} + \Delta t\boldsymbol{M}_{cm}^{EXT}(t+\phi\Delta t)\right), \tag{2.40}$$

where

$$\boldsymbol{M}_{cm}^{EXT}(t+\phi\Delta t) \approx \phi\boldsymbol{M}_{cm}^{EXT}(t+\Delta t) + (1-\phi)\boldsymbol{M}_{cm}^{EXT}(t), \tag{2.41}$$

which yields an implicit nonlinear equation, of the form $\boldsymbol{\omega}(t+\Delta t) = \boldsymbol{\mathcal{F}}(\boldsymbol{\omega}(t+\Delta t))$ since the moment of inertia is a function of time, $\overline{\boldsymbol{\mathcal{I}}}(t+\Delta t)$, due to the body's rotation. An iterative, implicit, solution scheme may be written as (for iterations $K = 1, 2, ...$)

$$\boldsymbol{\omega}^{K+1}(t+\Delta t) = \left(\overline{\boldsymbol{\mathcal{I}}}^K(t+\Delta t)\right)^{-1} \cdot \left((\overline{\boldsymbol{\mathcal{I}}} \cdot \boldsymbol{\omega})|_t + \Delta t \boldsymbol{M}_{cm}^{EXT,K}(t+\phi\Delta t)\right), \quad (2.42)$$

where $\overline{\boldsymbol{\mathcal{I}}}^K(t+\Delta t)$ can be computed by a similarity transform (described shortly).[3] After the update for $\boldsymbol{\omega}^{K+1}(t+\Delta t)$ has been computed (utilizing the $\overline{\boldsymbol{\mathcal{I}}}^K(t+\Delta t)$ from the previous iteration), the rotation of the body about the center of mass can be determined. The *incremental* angular rotation around the instantaneous rotation axis $\boldsymbol{a}^{K+1}(t+\phi\Delta t)$ (which will also have to be updated) is obtained by ($\boldsymbol{\omega}^{K+1}(t+\phi\Delta t) = \omega^{K+1}(t+\phi\Delta t)\boldsymbol{a}^{K+1}(t+\phi\Delta t)$)

$$\frac{d\theta^{K+1}}{dt}(t+\phi\Delta t) = \omega^{K+1}(t+\phi\Delta t) \approx \frac{\Delta\theta^{K+1}(t+\phi\Delta t)}{\Delta t}, \quad (2.43)$$

where $\omega^{K+1}(t+\phi\Delta t) = ||\boldsymbol{\omega}^{K+1}(t+\phi\Delta t)||$ being *a scalar* rotation rate about the instantaneous axis ($\Delta\theta^{K+1}(t+\phi\Delta t)$ is the corresponding incremental rotation about that instantaneous axis),

$$\boldsymbol{a}^{K+1}(t+\phi\Delta t) \stackrel{\text{def}}{=} \frac{\boldsymbol{\omega}^{K+1}(t+\phi\Delta t)}{||\boldsymbol{\omega}^{K+1}(t+\phi\Delta t)||} \approx \frac{\phi\boldsymbol{\omega}^{K+1}(t+\Delta t) + (1-\phi)\boldsymbol{\omega}(t)}{||\phi\boldsymbol{\omega}^{K+1}(t+\Delta t) + (1-\phi)\boldsymbol{\omega}(t)||}, \quad (2.44)$$

and thus

$$\Delta\theta^{K+1}(t+\phi\Delta t) = \omega^{K+1}(t+\phi\Delta t)\Delta t, \quad (2.45)$$

where $\omega^{K+1}(t+\phi\Delta t) = ||\phi\boldsymbol{\omega}^{K+1}(t+\Delta t) + (1-\phi)\boldsymbol{\omega}(t)||$. To determine the movement of the individual points/particles in the rigid (cluster) body, we need to perform a rigid body translation and rotation (described in the next section). For example, consider a point $\boldsymbol{r}_i$ on the body. The update would be

$$\boldsymbol{r}_i(t+\Delta t) = \boldsymbol{r}_i(t) + \underbrace{\boldsymbol{u}_{cm}}_{due\ to\ cm\ translation} + \underbrace{\boldsymbol{u}_{i,rot}}_{due\ to\ rotation\ wrt\ cm} \quad (2.46)$$

where

$$\boldsymbol{u}_{cm} = \boldsymbol{r}_{cm}(t+\Delta t) - \boldsymbol{r}_{cm}(t), \quad (2.47)$$

and where $\boldsymbol{u}_{i,rot}$ is a contribution due to an incremental rotation of the relative position vector

$$\boldsymbol{\tau}^{(i)} \stackrel{\text{def}}{=} \boldsymbol{r}_i(t) - \boldsymbol{r}_{cm}(t) \quad (2.48)$$

by $\Delta\theta$ about the center of mass (Fig. 2.3).

---

[3] One may view the overall process as a fixed-point calculation of the form $\boldsymbol{\omega}^{K+1}(t+\Delta t) = \mathcal{F}(\boldsymbol{\omega}^K(t+\Delta t))$.

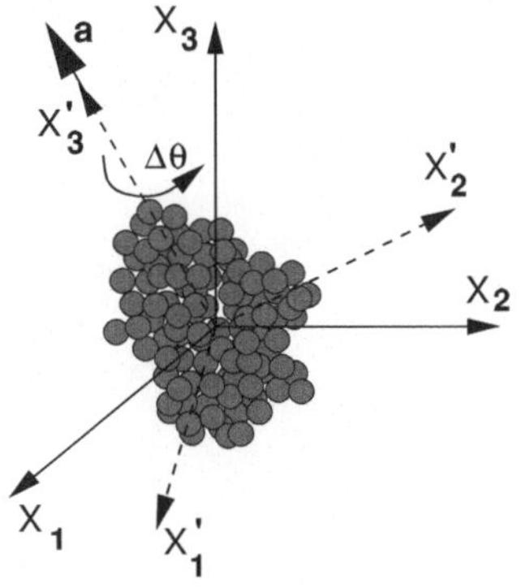

Fig. 2.3 Aligning the primed coordinate system with the instantaneous axis of rotation (*a*) for a cluster (Zohdi [44])

2.3.3 Transformation Matrices for Updates and Incremental Rotation

In order to rotate any point i, with position vector $\boldsymbol{\tau}^{(i)}$, associated with the rigid body, we require some standard transformations. The same transformation is needed to rotate the body's moment of inertia, $\overline{\mathcal{I}}$ (Fig. 2.3). It is a relatively standard exercise in linear algebra to show that any vector, $\boldsymbol{\tau}$, which can be expressed in either the unprimed or primed basis, $\boldsymbol{\tau} = (\boldsymbol{\tau} \cdot \boldsymbol{e}_i)\boldsymbol{e}_i = (\boldsymbol{\tau} \cdot \boldsymbol{e}'_j)\boldsymbol{e}'_j$, where summation index notation is employed. These two representations are explicitly related by

$$\begin{bmatrix} \tau_1 \\ \tau_2 \\ \tau_3 \end{bmatrix}' = \underbrace{\begin{bmatrix} \boldsymbol{e}_1 \cdot \boldsymbol{e}'_1 & \boldsymbol{e}_2 \cdot \boldsymbol{e}'_1 & \boldsymbol{e}_3 \cdot \boldsymbol{e}'_1 \\ \boldsymbol{e}_1 \cdot \boldsymbol{e}'_2 & \boldsymbol{e}_2 \cdot \boldsymbol{e}'_2 & \boldsymbol{e}_3 \cdot \boldsymbol{e}'_2 \\ \boldsymbol{e}_1 \cdot \boldsymbol{e}'_3 & \boldsymbol{e}_2 \cdot \boldsymbol{e}'_3 & \boldsymbol{e}_3 \cdot \boldsymbol{e}'_3 \end{bmatrix}}_{[\boldsymbol{A}]} \begin{bmatrix} \tau_1 \\ \tau_2 \\ \tau_3 \end{bmatrix}. \tag{2.49}$$

Note that $\boldsymbol{A}^{-1} = \boldsymbol{A}^T$, and thus $\boldsymbol{\tau}' = \boldsymbol{A} \cdot \boldsymbol{\tau}$ and $\boldsymbol{\tau} = \boldsymbol{A}^T \cdot \boldsymbol{\tau}'$. This basic result can be used to perform rotation of a vector about an axis, as well as the rotation of the inertia tensor. Without any loss of generality, we align the $\boldsymbol{e}'_3$ axis to instantaneous rotation axis $\boldsymbol{a}$. The total transformation (rotation) of a vector $\boldsymbol{\tau}^{(i)}$ representing a point i on the body can be represented by

$$[\boldsymbol{\tau}^{(i)}]^{rot} = \underbrace{[\boldsymbol{A}]^T \underbrace{[\boldsymbol{R}(\Delta\theta)] \underbrace{[\boldsymbol{A}][\boldsymbol{\tau}_i]}_{[\boldsymbol{\tau}^{(i)}]'}}_{[\boldsymbol{\tau}^{(i)}]^{rot,\prime}}}_{[\boldsymbol{\tau}^{(i)}]^{rot}} \tag{2.50}$$

where

$$[\boldsymbol{R}(\Delta\theta)] = \begin{bmatrix} cos(\Delta\theta) & -sin(\Delta\theta) & 0 \\ sin(\Delta\theta) & cos(\Delta\theta) & 0 \\ 0 & 0 & 1 \end{bmatrix}. \tag{2.51}$$

Similarly, for the rotation inertia tensor,

$$[\overline{\mathcal{I}}]^{rot} = \underbrace{[\boldsymbol{A}]^T \underbrace{[\boldsymbol{R}(\Delta\theta)] \underbrace{[\boldsymbol{A}][\overline{\mathcal{I}}][\boldsymbol{A}]^T}_{[\overline{\mathcal{I}}]'} [\boldsymbol{R}(\Delta\theta)]^T}_{[\overline{\mathcal{I}}]^{rot,\prime}} [\boldsymbol{A}]}_{[\overline{\mathcal{I}}]^{rot}}, \tag{2.52}$$

where, during the iterative calculations, $[\overline{\mathcal{I}}] = [\overline{\mathcal{I}}(t)]$ and $[\overline{\mathcal{I}}]^{rot} = [\overline{\mathcal{I}}(t + \Delta t))]$.

2.3.4 Algorithmic Procedure

The overall procedure is as follows, at time t:

1. Compute the new position of the center of mass.
2. Compute (iteratively) the incremental angular rotation of the body with respect to the center of mass until system convergence:

$$||\boldsymbol{\omega}^{K+1}(t + \Delta t) - \boldsymbol{\omega}^{K}(t + \Delta t)|| \leq TOL||\boldsymbol{\omega}^{K+1}(t + \Delta t)||. \tag{2.53}$$

This requires a rotation of the body within the iterations:

(a) Update $\boldsymbol{\omega}^{K+1}(t + \Delta t)$ (assuming that $\boldsymbol{\omega}^{K}(t + \Delta t)$ has been computed)

$$\boldsymbol{\omega}^{K+1}(t+\Delta t) = \left(\overline{\mathcal{I}}^{K}(t + \Delta t)\right)^{-1} \cdot \left((\overline{\mathcal{I}} \cdot \boldsymbol{\omega})|_t + \Delta t \boldsymbol{M}_{cm}^{EXT,K}(t + \phi\Delta t)\right). \tag{2.54}$$

(b) Compute the (updated) axis of rotation:

$$\boldsymbol{a}^{K+1}(t + \phi\Delta t) \stackrel{\text{def}}{=} \frac{\boldsymbol{\omega}^{K+1}(t + \phi\Delta t)}{||\boldsymbol{\omega}^{K+1}(t + \phi\Delta t)||} \approx \frac{\phi\boldsymbol{\omega}^{K+1}(t + \Delta t) + (1 - \phi)\boldsymbol{\omega}(t)}{||\phi\boldsymbol{\omega}^{K+1}(t + \Delta t) + (1 - \phi)\boldsymbol{\omega}(t)||}. \tag{2.55}$$

(c) Compute the basis $\boldsymbol{e}_3'$-aligned instantaneous axis of rotation ($\boldsymbol{a}$):
   (i) $\boldsymbol{e}_3'$, is aligned with $\boldsymbol{a}^{K+1}$
   (ii) $\boldsymbol{e}_1' = \boldsymbol{e}_3' \times \boldsymbol{e}_3/||\boldsymbol{e}_3' \times \boldsymbol{e}_3||$ and
   (iii) $\boldsymbol{e}_2' = \boldsymbol{e}_3' \times \boldsymbol{e}_1'/||\boldsymbol{e}_3' \times \boldsymbol{e}_1'||$.

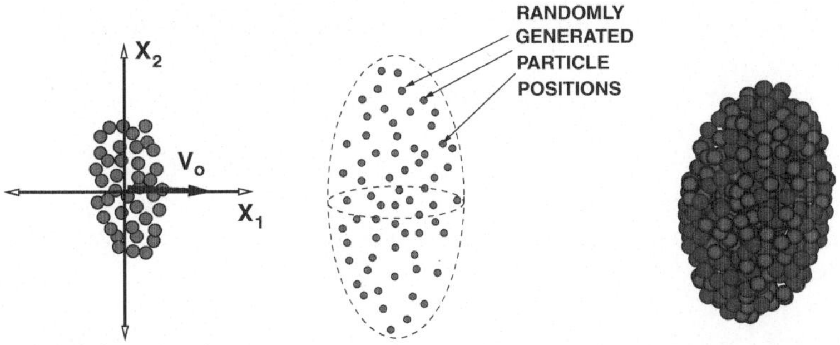

Fig. 2.4 *Left* the initial configuration. *Middle* an ellipsoidal envelope for the random particle positions. *Right* the actual object used in computations (*blue* is a base-positive charge and *red* is a base-negative charge) (Zohdi [44])

(d) Compute incremental angle of rotation $\Delta\theta^{K+1}(t+\phi\Delta t)$ and the composite transformation for the inertia tensor in Eq. 2.52 and obtain the update $\overline{\mathcal{I}}^{K+1}(t+\Delta t)$.
(e) Compute the total new position of the points in the body (i) with Eqs. 2.50 and 2.46.
(f) Repeat steps (a)–(e) until Eq. 2.53 is satisfied.

3. Increment time forward and repeat the procedure.

2.4 Model Problems/Numerical Examples

As a model problem, we consider a cluster formed by randomly dispersing charged particulates within a prolate ellipsoidal domain (aspect ratio of 2:1, Fig. 2.4). The radii of the ellipsoidal domain (envelope) in which the particles were randomly dispersed were 0.002 m (major axis) and 0.001 m (for both minor axes).[4] We considered $N_c = 500$ randomly distributed particles, with an overall charge of the body set to $\sum_{i=1}^{N_c} q_i = q^* = 0.01$ Coulomb, where for $q_i = \pm q_o + q^*/N_c$. There were 250 base-positive ($q_o = +0.001$) and 250 base-negative ($q_o = -0.001$) particles in the system. The radii of the individual particles were set to $r = 0.0001$ m.[5]

In order to help investigate what type of motion a charged cluster will experience, we consider a "comparison" case when only a single charged particle (or, considered equivalently, a lumped charged mass) is present in the system, with position vector

---

[4] The absolute length-scales, charges and masses are somewhat irrelevant to the model problem framework, and can be scaled to any desired value for specific application.

[5] The densities for the particles were uniformly assigned $\rho = 2000$ kg/m^3, with masses given by $m = \rho\frac{4}{3}\pi R^3$ kg.

denoted by $\boldsymbol{r}_s$, governed by Eq. 2.1, written in a slightly different form as ($\dot{\boldsymbol{r}}_s = \boldsymbol{v}_s$)[6]

$$m_s \ddot{\boldsymbol{r}}_s = q_s(\boldsymbol{E}^{ext} + \dot{\boldsymbol{r}}_s \times \boldsymbol{B}^{ext}). \tag{2.56}$$

The difference in the solution path for the single particle (governed by Eq. 2.56) and a multiparticle cluster, with center of mass given by $\boldsymbol{r}_{cm}$ (governed by Eq. 2.26 and implicitly by 2.27), can be characterized by taking the difference between Eqs. 2.26 and 2.56 to obtain

$$\mathcal{M}\ddot{\boldsymbol{r}}_{cm} - m\ddot{\boldsymbol{r}}_s = (Q - q_s)\boldsymbol{E}^{ext} + (Q\dot{\boldsymbol{r}}_{cm} - q_s\dot{\boldsymbol{r}}_s) \times \boldsymbol{B}^{ext}) + (\boldsymbol{\omega} \times \boldsymbol{R}_q \times \boldsymbol{B}^{ext}). \tag{2.57}$$

If we assume $Q = q_s$, $M = m_s$ and define $\boldsymbol{\mathcal{E}} \stackrel{\text{def}}{=} \boldsymbol{r}_{cm} - \boldsymbol{r}_s$, we obtain a "deviation" equation governing the difference in the trajectories of the two systems, that is,

$$\ddot{\boldsymbol{\mathcal{E}}} = \frac{1}{\mathcal{M}} \left(Q\dot{\boldsymbol{\mathcal{E}}} \times \boldsymbol{B}^{ext} + \boldsymbol{\omega} \times \boldsymbol{R}_q \times \boldsymbol{B}^{ext} \right). \tag{2.58}$$

In the special case when $\boldsymbol{B}^{ext} = \boldsymbol{0}$, the difference between the motion of the center of mass is zero ($\boldsymbol{\mathcal{E}}(t) = 0$), although one can still expect rotation about the center of mass for the cluster, via $\boldsymbol{\omega}$, which is dictated by Eq. 2.27.

2.4.1 Special Case # 1: No Magnetic Field ($\boldsymbol{E}^{ext} \neq \boldsymbol{0}$ and $\boldsymbol{B}^{ext} = \boldsymbol{0}$)

In the special case when there is no magnetic field, if $\boldsymbol{r}_s(t=0) = \boldsymbol{0}$, $\boldsymbol{v}_s(t=0) = v_o\boldsymbol{e}_1$, $\boldsymbol{B}^{ext} = \boldsymbol{0}$ and $\boldsymbol{E}^{ext} = E^{ext}\boldsymbol{e}_3$, the solution for the dynamics of a single particle is

$$\begin{Bmatrix} v_{s1}(t) \\ v_{s2}(t) \\ v_{s3}(t) \end{Bmatrix} = \begin{Bmatrix} v_o \\ 0 \\ \frac{q_s}{m_s} E_3^{ext} t \end{Bmatrix} \Rightarrow \begin{Bmatrix} r_{s1}(t) \\ r_{s2}(t) \\ r_{s3}(t) \end{Bmatrix} = \begin{Bmatrix} v_o t \\ 0 \\ \frac{q_s}{2m_s} E_3^{ext} t^2 \end{Bmatrix}. \tag{2.59}$$

Now, for a cluster,[7] let us (numerically) consider this special case ($\boldsymbol{v}(t = 0) = 0.01\boldsymbol{e}_1$, $\boldsymbol{\omega}(t = 0) = \boldsymbol{0}$, $\boldsymbol{B}^{ext} = \boldsymbol{0}$ and $\boldsymbol{E}^{ext} = 0.1\boldsymbol{e}_3$), which should yield results similar to a single particle in Eq. 2.59. Indeed, as shown in Fig. 2.5, as in the single particle case, when $\boldsymbol{B}^{ext} = \boldsymbol{0}$, we have the predicted motion, i.e., the center of mass

[6] Note: The governing Eq. 2.56, written in component form is, for component 1: $\dot{v}_{s1} = \frac{q_s}{m_s}(E_1^{ext} + (v_{s2}B_3^{ext} - v_{s3}B_2^{ext}))$, for component 2: $\dot{v}_{s2} = \frac{q_s}{m_s}(E_2^{ext} - (v_{s1}B_3^{ext} - v_{s3}B_1^{ext}))$, and for component 3: $\dot{v}_{s3} = \frac{q_s}{m_s}(E_3^{ext} + (v_{s1}B_2^{ext} - v_{s2}B_1^{ext}))$. These equations can be solved analytically. There are a variety of possible particle trajectories, and we refer the reader to Jackson [13].

[7] These specific parameter choices resulted in $||\boldsymbol{R}_q||$=0.00297 and $||\boldsymbol{\mathcal{I}}_q||$= 0.0000229. The simulations were run for other large clusters with similar trends occurring. In other words, these results are representative. The overall approach is general, and is valid for any distribution of charges.

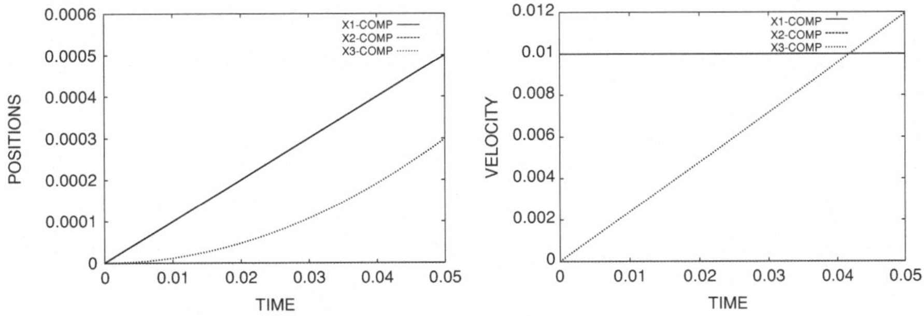

Fig. 2.5 Special case 1: $\boldsymbol{v}(t=0) = 0.01\boldsymbol{e}_1$, $\boldsymbol{B}^{ext} = \mathbf{0}$ and $\boldsymbol{E}^{ext} = 0.1\boldsymbol{e}_3$. *Left* the position of the center of mass. *Right* the velocity of the center of mass. The trajectory of the center of mass of the cluster is similar to that of a single charged particle (Zohdi [44])

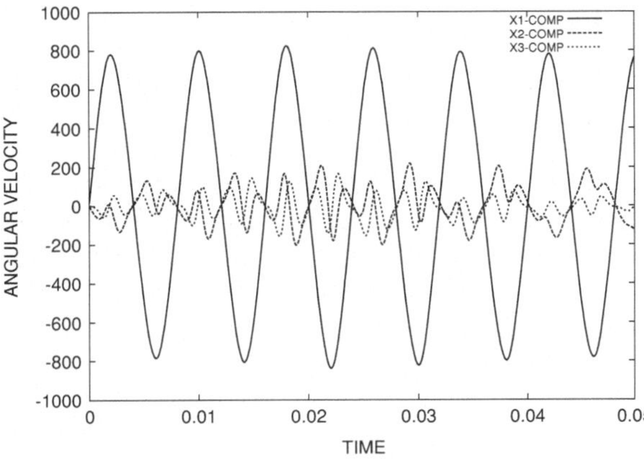

Fig. 2.6 Special case 1: $\boldsymbol{v}(t=0) = 0.01\boldsymbol{e}_1$, $\boldsymbol{B}^{ext} = \mathbf{0}$ and $\boldsymbol{E}^{ext} = 0.1\boldsymbol{e}_3$: the angular velocity. The rotation about the center of mass is induced entirely by the electric field (Zohdi [44])

of a cluster behaves similarly to that of a single particle. Notice in Fig. 2.6, because of the term $\mathcal{T}_4 \neq 0$ in Eq. 2.27, however, that there is rotational motion about the center of mass. This rotation (a twisting back and forth) about the center of mass is purely induced by the electric field. We emphasize, for this special case, because the only "forcing term" on the right hand side of the angular momentum Eq. 2.27 is $\mathcal{T}_4 \stackrel{\text{def}}{=} \boldsymbol{R}_q \times \boldsymbol{E}^{ext}$ (since $\boldsymbol{B}^{ext} = \mathbf{0}$), $\boldsymbol{E}^{ext}$ is solely responsible for any spin of the cluster about its mass center. This term manifests the oscillations seen in Fig. 2.6 and is a phenomenon that is nonexistent in the single particle solution Eq. 2.59.

2.4.2 Special Case # 2: No Electric Field ($\boldsymbol{E}^{ext} = \mathbf{0}$ and $\boldsymbol{B}^{ext} \neq \mathbf{0}$)

Now, consider a case with no electric field and a magnetic field present, $\boldsymbol{r}_s(t=0) = \mathbf{0}$, $\boldsymbol{v}_s(t=0) = v_o\boldsymbol{e}_1$, $\boldsymbol{B}^{ext} = B_3^{ext}\boldsymbol{e}_3$ and $\boldsymbol{E}^{ext} = \mathbf{0}$. Consequently, for a single particle, the solution is

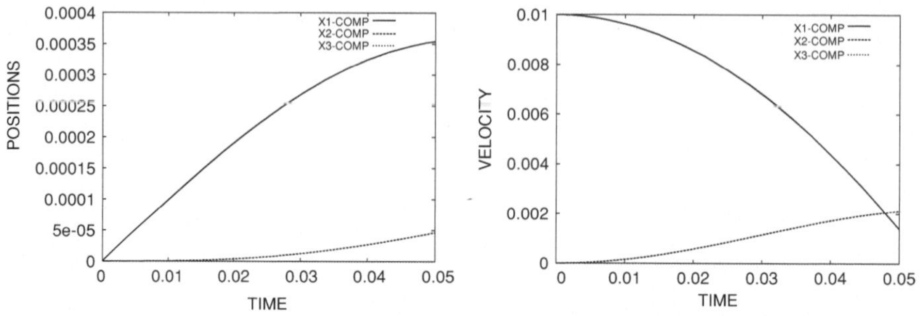

Fig. 2.7 Special case 2: $\boldsymbol{v}(t=0) = 0.01\boldsymbol{e}_1$, $\boldsymbol{B}^{ext} = 0.01\boldsymbol{e}_3$ and $\boldsymbol{E}^{ext} = \boldsymbol{0}$. *Left* the position of the center of mass. *Right* the velocity of the center of mass (Zohdi [44])

$$\begin{Bmatrix} v_{s1}(t) \\ v_{s2}(t) \\ v_{s3}(t) \end{Bmatrix} = \begin{Bmatrix} v_o cos\Omega_s t \\ -v_o sin\Omega_s t \\ 0 \end{Bmatrix} \Rightarrow \begin{Bmatrix} r_{s1}(t) \\ r_{s2}(t) \\ r_{s3}(t) \end{Bmatrix} = \begin{Bmatrix} \frac{v_o}{\Omega_s} sin\Omega_s t \\ \frac{v_o}{\Omega_s}(cos\Omega_s t - 1) \\ 0 \end{Bmatrix}, \tag{2.60}$$

where $\Omega_s = \frac{q_s B_3^{ext}}{m_s}$ is known as the cyclotron frequency. The cyclotron frequency (gyrofrequency) is the angular frequency at which a charged particle makes circular orbits in a plane perpendicular to the static magnetic field. Notice that when $E_3^{ext} = 0$, this traces out the equation of a circle centered at $(0, \frac{-v_o}{\Omega_s}, 0)$. The radius of the "magnetically-induced circle" (radius of oscillation) is[8]

$$\mathcal{R} \stackrel{\text{def}}{=} \frac{v_o}{\Omega_s} = \frac{v_o m_s}{q_s B_3^{ext}}. \tag{2.61}$$

Thus, if a desired "turning radius" is denoted by $\mathcal{R}$, one may solve for the magnetic field that delivers the desired effect, $B_3^{ext} = \frac{v_o m_s}{q_s \mathcal{R}}$. We define the corresponding time period for one cycle to be completed as $T \stackrel{\text{def}}{=} 2\pi/\Omega_s$. For the parameters chosen, this results in $\mathcal{R} = 0.0004188$ m, $\Omega_s = 23.87$ rad/s and $T = 0.1315$ s. For a cluster (numerically computed with an initial $\boldsymbol{\omega}(t=0)$), Fig. 2.7 illustrates motion with a possible long range period T and "large-scale" turning (cyclotron) radius ($\mathcal{R}$) that is similar to that of a single particle.[9] Figure 2.8 indicates that there is some slight rotation of the body around the center of mass.

---

[8] As mentioned earlier, this field generates helical motion in three dimensions when $\boldsymbol{E}^{ext} \neq \boldsymbol{0}$.

[9] Figure 2.7 illustrates approximately one-quarter of a period (total period $T \approx 20$ s) and a cyclotron radius of $\mathcal{R} \approx 0.0004$ m.

Fig. 2.8 Special case 2: $\boldsymbol{v}(t=0)=0.01\boldsymbol{e}_1$, $\boldsymbol{B}^{ext}=0.01\boldsymbol{e}_3$ and $\boldsymbol{E}^{ext}=\boldsymbol{0}$: the angular velocity (Zohdi [44])

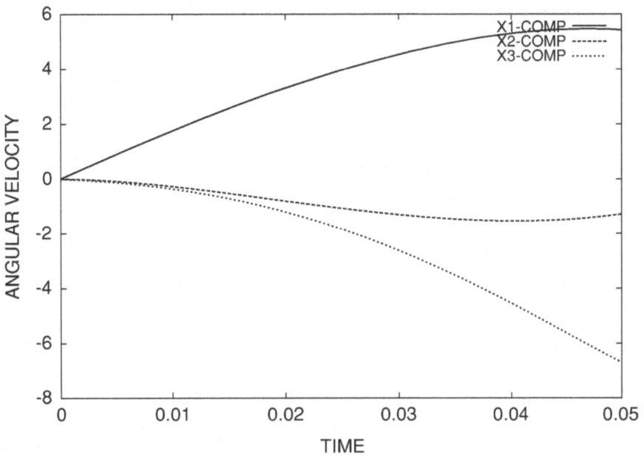

2.4.3 General Case # 3: Combined Electric and Magnetic Fields ($\boldsymbol{E}^{ext} \neq \boldsymbol{0}$ and $\boldsymbol{B}^{ext} \neq \boldsymbol{0}$)

Now, consider both the electric and magnetic fields to be present, $\boldsymbol{r}_s(t=0)=\boldsymbol{0}$, $\boldsymbol{v}_s(t=0)=v_o\boldsymbol{e}_1$, $\boldsymbol{B}^{ext}=B_3^{ext}\boldsymbol{e}_3$ and $\boldsymbol{E}^{ext}=E_3^{ext}\boldsymbol{e}_3$, consequently, for a single particle

$$\begin{Bmatrix} v_{s1}(t) \\ v_{s2}(t) \\ v_{s3}(t) \end{Bmatrix} = \begin{Bmatrix} v_o cos\Omega_s t \\ -v_o sin\Omega_s t \\ \frac{q_s}{m_s}E_3^{ext}t \end{Bmatrix} \Rightarrow \begin{Bmatrix} r_{s1}(t) \\ r_{s2}(t) \\ r_{s3}(t) \end{Bmatrix} = \begin{Bmatrix} \frac{v_o}{\Omega_s} sin\Omega_s t \\ \frac{v_o}{\Omega_s}(cos\Omega_s t - 1) \\ \frac{q_s}{2m_s}E_3^{ext}t^2 \end{Bmatrix}. \tag{2.62}$$

Let us now consider these parameters for a cluster ($\boldsymbol{r}(t=0)=\boldsymbol{0}$, $\boldsymbol{\omega}(t=0)=\boldsymbol{0}$, $\boldsymbol{v}(t=0)=0.01\boldsymbol{e}_1$, $\boldsymbol{B}^{ext}=0.01\boldsymbol{e}_3$ and $\boldsymbol{E}^{ext}=0.1\boldsymbol{e}_3$). As shown in Fig. 2.9, there is large-scale reversal of the x_1 component with a superposed oscillatory "wobble." As in special case # 2, there would be large-scale turning of the cluster, albeit slowly-induced, due to a (nonmonotonic) reversal of the x_1 velocity, which would eventually trace out a helix-like path in the $x_1 - x_2$ plane moving upwards in the x_3 direction. The rotations about the center of mass are highly variable due to the random positions of the charged particles within the cluster. Figures 2.9 and 2.10 indicates the dramatic difference (due to the absence of the electric field) between special case # 2 and general case # 3, which is predicted by Eq. 2.58. The key observation is that the effects of $\boldsymbol{E}^{ext}$ and $\boldsymbol{B}^{ext}$ are strongly coupled via Eqs. 2.26 and 2.27, as opposed to uncoupled (as exhibited by Eqs. 2.60 and 2.62). One reason for this strong coupling is due to the dependence of $\boldsymbol{\omega}$ on $\boldsymbol{E}^{ext}$, as exhibited by Eq. 2.27. Even when $Q = q_s = 0$, the trajectory deviation is governed by

$$\ddot{\boldsymbol{\mathcal{E}}} = \frac{1}{\mathcal{M}}\left(\boldsymbol{\omega} \times \boldsymbol{R}_q \times \boldsymbol{B}^{ext}\right), \tag{2.63}$$

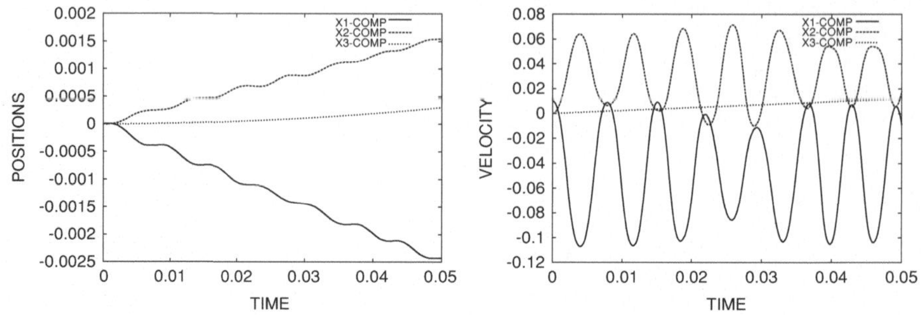

Fig. 2.9 General case 3: $\boldsymbol{v}(t = 0) = 0.01\boldsymbol{e}_1$, $\boldsymbol{B}^{ext} = 0.01\boldsymbol{e}_3$ and $\boldsymbol{E}^{ext} = 0.1\boldsymbol{e}_3$. *Left* the position of the center of mass. *Right* the velocity of the center of mass (Zohdi [44])

and is strongly influenced by a balance of angular momentum, through $\boldsymbol{\omega}$, which is governed by Eq. 2.27. The influence of $\boldsymbol{E}^{ext}$ comes through Eq. 2.27, via term $\mathcal{T}_4$, even when $Q = 0$. We further note that in the special case when $\boldsymbol{R}_q = \boldsymbol{0}$ and $Q = q_s \neq 0$, the deviation is governed by

$$\ddot{\mathcal{E}} = \frac{1}{\mathcal{M}} \left(Q\dot{\mathcal{E}} \times \boldsymbol{B}^{ext} \right), \tag{2.64}$$

and the magnetic field plays a strong role in the deviation of the trajectories. One can still expect rotation about the center of mass for the cluster due to term $\mathcal{T}_6$, which is dependent on $\overline{\mathcal{I}}_q$, in Eq. 2.27, even when $\boldsymbol{R}_q = \boldsymbol{0}$.

2.5 Closing Remarks

It was shown that Eqs. 2.26 and 2.27 govern the dynamics of a charged cluster, and that the following three quantities play a central role in the cluster behavior:

- the sum of the individual (q_i) charges: $Q \stackrel{\text{def}}{=} \sum_{i=1}^{N_c} q_i$ ("overall charge" or the first moment),
- the sum of the distances between the individual charged particles ($\boldsymbol{r}_i$) and the center of mass of the cluster ($\boldsymbol{r}_{cm}$), weighted by the individual charges: $\boldsymbol{R}_q \stackrel{\text{def}}{=} \sum_{i=1}^{N_p} q_i(\boldsymbol{r}_i - \boldsymbol{r}_{cm})$ ("charged radius" or the second-moment) and
- the sum of the self-cross-product of the distances between the individual charged particles and the center of mass of the cluster, weighted by the individual charges ($\overline{\mathcal{I}}_q$): $\boldsymbol{H}_q = \overline{\mathcal{I}}_q \cdot \boldsymbol{\omega} \stackrel{\text{def}}{=} \sum_{i=1}^{N_p} q_i(\boldsymbol{r}_i - \boldsymbol{r}_{cm}) \times \boldsymbol{\omega} \times (\boldsymbol{r}_i - \boldsymbol{r}_{cm})$, where $\boldsymbol{\omega}$ is the angular velocity of the body and $\overline{\mathcal{I}}_q$ has components of $(p, s = 1, 2, 3)$ $\bar{\mathcal{I}}_{q,ps} = \sum_{i=1}^{N_c} q_i(||\boldsymbol{r}_i - \boldsymbol{r}_{cm}||^2 \delta_{ps} - \hat{x}_{ip}\hat{x}_{is})$ ("moment of charge" or the third-moment).

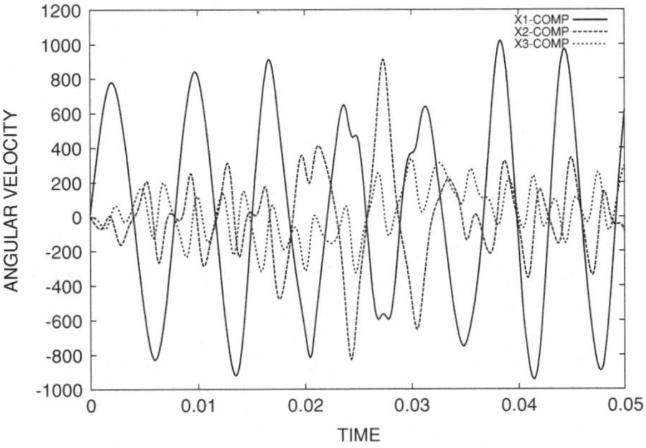

Fig. 2.10 General case 3: $\boldsymbol{v}(t=0) = 0.01\boldsymbol{e}_1$, $\boldsymbol{B}^{ext} = 0.01\boldsymbol{e}_3$ and $\boldsymbol{E}^{ext} = 0.1\boldsymbol{e}_3$: the angular velocity (Zohdi [44])

In addition to dictating the motion of a cluster, these quantities control the differences in the motion of a charged cluster relative to that of a single charged particle. The deviation in dynamics of a charged cluster and a single charged particle (or lumped charged mass) is strongly influenced by the simultaneous presence of an electric and magnetic field. In particular, for large off-center charges, characterized by $\boldsymbol{R}_q$ and $\overline{\mathcal{I}}_q$, the motion will vary significantly, and is governed by Eq. 2.58.

Independent of the purely scientific interest in the dynamics of a charged cluster, there are implications of these results for large-scale computation of particulate flows. Within the last decade, simultaneous advances in computational methods, applied mathematics and high-performance computing has raised the possibility that an analyst can directly numerically simulate (DNS) a process employing particulate flows containing several million particles, incorporating all of the important microscale details. A relatively straightforward DNS-type formulation of the dynamics of a multiparticulate system is to track the motion of $i = 1, 2 ... N$ particles,

$$m_i \ddot{\boldsymbol{r}}_i = \boldsymbol{\Psi}_i^{tot}(\boldsymbol{r}_1, \boldsymbol{r}_2, ..., \boldsymbol{r}_{N_p}) = \boldsymbol{\Psi}_i^{nf} + \boldsymbol{\Psi}_i^{con} + q_i(\boldsymbol{E}^{ext} + \boldsymbol{v}_i \times \boldsymbol{B}^{ext}), \quad (2.65)$$

where $\boldsymbol{r}_i$ is the position vector of the ith particle, $\boldsymbol{\Psi}_i^{tot}$ represents all forces acting on particle i, $\boldsymbol{\Psi}_i^{nf}$ represents near-field inter-(charged)particle forces acting on particle i, $\boldsymbol{\Psi}_i^{con}$ represents contact forces acting on particle i and where $\boldsymbol{E}^{ext}$ and $\boldsymbol{B}^{ext}$ are the external electromagnetic fields. The simulation of such flowing particulate systems has been extensively investigated for the last decade by Zohdi [27–45], employing numerical schemes based on high-performance iterative solvers, sorting-binning for fast interparticle calculations, Verlet lists, domain decomposition, parallel processing and temporally-adaptive methods. These types of formulations can easily describe the interaction of multiple particulate jets, jet breakup/disintegration and jet impact, where the application of continuum approaches is complex. The dynamics of clusters that evolve within a jet can be represented by directly describing the motion of each individual particle with the appropriate binding conditions (constraints) to its neighbors in the cluster, and is discussed next.

References

1. Arbelaez, D., Zohdi, T. I., & Dornfeld, D. (2008). Modeling and simulation of material removal with particulate flow. *Computational Mechanics, 42*, 749–759.
2. Arbelaez, D., Zohdi, T. I., & Dornfeld, D. (2009). On impinging near-field granular jets. *International Journal for Numerical Methods in Fluids.*
3. Ciampini, D., Spelt, J. K., & Papini, M. (2003). Simulation of interference effects in particle streams following impact with a flat surface. Part I. Theory and analysis. *Wear, 254*, 237–249.
4. Ciampini, D., Spelt, J. K., & Papini, M. (2003). Simulation of interference effects in particle streams following impact with a flat surface. Part II. Parametric study and implications for erosion testing and blast cleaning. *Wear, 254*, 250–264.
5. Fish, J. (2006). Bridging the scales in nano engineering and science. *Journal of Nanoparticle Research, 8*, 577–594.
6. Frenklach, M., & Carmer, C. S. (1999). Molecular dynamics using combined quantum & empirical forces: Application to surface reactions. *Advances in Classical Trajectory Methods, 4*, 27–63.
7. Ghobeity, A., Krajac, T., Burzynski, T., Papini, M., & Spelt, J. K. (2008). Surface evolution models in abrasive jet micromachining. *Wear, 264*, 185–198.
8. Ghobeity, A., Spelt, J. K., & Papini, M. (2008). Abrasive jet micro-machining of planar areas and transitional slopes. *Journal of Micromechanics and Microengineering, 18*, 055014.
9. Ghosh, S., Valiveti, D. M., Harris, S. H., & Boileau, J. (2006). Microstructure characterization based domain partitioning as a pre-processor to multi-scale modeling of cast Aluminum alloys. *Modelling and Simulation in Material Science and Engineering, 14*, 1363–1396.
10. Gomes-Ferreira, C., Ciampini, D., & Papini, M. (2004). The effect of inter-particle collisions in erosive streams on the distribution of energy flux incident to a flat surface. *Tribology International, 37*, 791–807.
11. Haile, J. M. (1992). *Molecular dynamics simulations: Elementary methods*. New York: Wiley.
12. Hase, W. L. (1999). *Molecular dynamics of clusters, surfaces, liquids, and interfaces.* Advances in classical trajectory methods. (Vol. 4). Stamford: JAI Press.
13. Jackson, J. D. (1998). *Classical electrodynamics* (3rd ed.). New York: Wiley.
14. Luo, L., & Dornfeld, D. A. (2001). Material removal mechanism in chemical mechanical polishing: theory and modeling. *IEEE Transactions on Semiconductor Manufacturing, 14*(2), 112–133.
15. Luo, L., & Dornfeld, D. A. (2003). Material removal regions in chemical mechanical planarization for sub-micron integration for sub-micron integrated circuit fabrication: coupling effects of slurry chemicals, abrasive size distribution, and wafer-pad contact area. *IEEE Transactions on Semiconductor Manufacturing,16*, 45–56.
16. Luo, L., & Dornfeld, D. A. (2003). Effects of abrasive size distribution in chemical-mechanical planarization: modeling and verification. *IEEE Transactions on Semiconductor Manufacturing, 16*, 469–476.
17. Luo, L., & Dornfeld, D. A. (2004). *Integrated modeling of chemical mechanical planarization of sub-micron IC fabrication*. New York: Springer.
18. Moelwyn-Hughes, E. A. (1961). *Physical Chemistry*. New York: Pergamon.
19. Rapaport, D. C. (1995). *The Art of Molecular Dynamics Simulation*. Cambridge: Cambridge University Press.
20. Rechtsman, M., Stillinger, F. H., & Torquato, S. (2005). Optimized interactions for targeted self-assembly: Application to honeycomb lattice. *Physical Review Letters, 95*, 228301.
21. Rechtsman, M., Stillinger, F. H., & Torquato, S. (2006). Designed interaction potentials via inverse methods for self-assembly. *Physical Review E, 73*, 011406.
22. Schlick, T. (2000). *Molecular modeling & simulation: An interdisciplinary guide*. New York: Springer.
23. Stillinger, F. H., & Weber, T. A. (1985). Computer simulation of local order in condensed phases of silicon. *Physical Review B, 31*, 5262–5271.

24. Tersoff, J. (1988). Empirical interatomic potential for carbon, with applications to amorphous carbon. *Physical Review Letters, 61*, 2879–2882.
25. Torquato, S. (2002). *Random heterogeneous materials: Microstructure and macroscopic properties*. New York: Springer.
26. Torquato, S. (2009). Inverse optimization techniques for targeted self-assembly. *Soft Matter, 5*, 1157.
27. Zohdi, T. I. (2002). An adaptive-recursive staggering strategy for simulating multifield coupled processes in microheterogeneous solids. *Internatioanl Journal for Numerical Methods in Engineering, 53*, 1511–1532.
28. Zohdi, T. I. (2003). Genetic design of solids possessing a random-particulate microstructure. *PTRS: Mathematical Physical and Engineering Sciences, 361*(1806), 1021–1043.
29. Zohdi, T. I. (2003). On the compaction of cohesive hyperelastic granules at finite strains. *Proceedings of the Royal Society, 454*(2034), 1395–1401.
30. Zohdi, T. I. (2003). Computational design of swarms. *Internatioanl Journal for Numerical Methods and Engineering, 57*, 2205–2219.
31. Zohdi, T. I. (2003). Constrained inverse formulations in random material design. *Computational Methods in Applied Mechanics and Engineering. 1–20 192*(28–30), 18, 3179–3194.
32. Zohdi, T. I. (2004). Modeling and simulation of a class of coupled thermo-chemo-mechanical processes in multiphase solids. *Computational Methods in, Applied Mechanics and Engineering, 193*(6–8), 679–699.
33. Zohdi, T. I. (2004). Modeling and direct simulation of near-field granular flows. *The International Journal of Solids Structures, 42*(2), 539–564.
34. Zohdi, T. I. (2004). A computational framework for agglomeration in thermo-chemically reacting granular flows. *Proceedings of the Royal Society, 460*(2052), 3421–3445.
35. Zohdi, T. I. (2005). Charge-induced clustering in multifield particulate flow. *International Journal of Numerical Methods and Engineering, 62*(7), 870–898.
36. Zohdi, T. I. (2006). Computation of the coupled thermo-optical scattering properties of random particulate systems. *Computational Methods in Applied Mechanics and Engineering, 195*, 5813–5830.
37. Zohdi, T. I. (2007). Computation of strongly coupled multifield interaction in particle-fluid systems. *Computational Methods in Applied Mechanics and Engrineering, 196*, 3927–3950.
38. Zohdi, T. I. (2007). Particle collision and adhesion under the influence of near-fields. *Journal of Mechanics of Materials and Structures, 2*(6), 1011–1018.
39. Zohdi, T. I. (2007). Introduction to the modeling and simulation of particulate flows. *Society for Industrial and Applied Mathematics*.
40. Zohdi, T. I. (2008). On the computation of the coupled thermo-electromagnetic response of continua with particulate microstructure. *International Journal of Numerical Methods in Engineering, 76*, 1250–1279.
41. Zohdi, T. I. (2009). Mechanistic modeling of swarms. *Computational Methods in Applied Mechanics and Engineering, 198*(21–26), 2039–2051.
42. Zohdi, T. I. (2010). Charged wall-growth in channel-flow. *International Journal of Engineering Sciences, 48*, 15–20.
43. Zohdi, T. I. (2010). On the dynamics of charged electromagnetic particulate jets. *Archieves of Computational Methods in Engineering, 17*(2), 109–135.
44. Zohdi, T. I. (2011). Dynamics of clusters of charged particulates in electromagnetic fields. *International Journal of Numerical Methods in Engineering, 85*, 1140–1159.
45. Zohdi, T. I., & Wriggers, P. (2008). *Introduction to computational micromechanics*, Second reprinting. New York: Springer.

Chapter 3
Dynamics of Flowing Charged Particles

We now consider the next level of complexity beyond rigid clusters of particles, namely, free-flowing systems of charged particles. Furthermore, we consider thermal effects in such systems, which partially arise due to interparticle impact, as well as contact with the external environment, i.e., obstacles.

3.1 Multiple Particulate Flow in the Presence of Near-Fields

The objects in the flow are assumed to be small enough to be considered (idealized) as spherical particles and that the effects of their rotation with respect to their center of mass is unimportant to their overall motion.

3.1.1 Particulate Dynamics with Near-Fields

We consider a group of nonintersecting particles (N_p in total). The approach in this section draws from methods developed in Zohdi [101, 103, 106]. The equation of motion for the ith particle in a flow is

$$m_i \ddot{\boldsymbol{r}}_i = \boldsymbol{\Psi}_i^{tot}(\boldsymbol{r}_1, \boldsymbol{r}_2, ..., \boldsymbol{r}_{N_p}), \tag{3.1}$$

where $\boldsymbol{r}_i$ is the position vector of the ith particle and where $\boldsymbol{\Psi}_i^{tot}$ represents all forces acting on particle i. Specifically,

$$\boldsymbol{\Psi}_i^{tot} = \boldsymbol{\Psi}_i^{nf} + \boldsymbol{\Psi}_i^{con} + \boldsymbol{\Psi}_i^{fric} + \boldsymbol{\Psi}_i^{env} \tag{3.2}$$

T. I. Zohdi, *Dynamics of Charged Particulate Systems*, SpringerBriefs in Applied Sciences and Technology, DOI: 10.1007/978-3-642-28519-6_3,

represents the sum of forces due to near-field interaction ($\boldsymbol{\Psi}^{nf}$, introduced in Eq. 1.51), normal contact impulse forces ($\boldsymbol{\Psi}^{con}$), frictional impulse forces ($\boldsymbol{\Psi}^{fric}$) and the surrounding environment ($\boldsymbol{\Psi}^{env}$).[1]

Remarks In many applications, the near-fields can dramatically change when the particles are very close to one another, leading to increased repulsion or attraction. A particularly straightforward way to model this is via a near-field attractive/repulsive augmentation of the form

$$\tilde{\boldsymbol{\Psi}}_i^{nf} = \boldsymbol{\Psi}_i^{nf} + \underbrace{\alpha_a ||\boldsymbol{r}_i - \boldsymbol{r}_j||^{\beta_a} \boldsymbol{n}_{ij}}_{\boldsymbol{\Psi}^a \stackrel{\text{def}}{=} \text{augmentation force}}, \tag{3.3}$$

which is activated if

$$||\boldsymbol{r}_i - \boldsymbol{r}_j|| \leq (b_i + b_j)\delta_a, \tag{3.4}$$

where b_i and b_j are the radii of the particles, and where $\delta_a \geq 0$ is the critical distance needed for the augmentation to become active. When $\delta_a = 1$ and $\alpha_a < 0$, then the model can be interpreted as a (contact) penalty for particle interpenetration. Many such "augmentation" models exist, for example, see Duran [20]. For many engineering materials, some surface adhesion persists, which can lead to bonding phenomena between surfaces, even when no explicit external charging has occurred. For example, see Tabor [85] and Rietema [78] (specifically for agglomeration).

3.1.2 Mechanical Contact with Near-Field Interaction

We now consider cases where mechanical contact occurs between particles, in the presence of near-field interaction. A primary simplifying assumption is made: *the particles remain spherical after impact, i.e., any permanent deformation is considered negligible.* For two colliding particles, i and j, normal to the line of impact, a balance of linear momentum relating the states before impact (time $= t$) and after impact (time $= t + \delta t$) reads as

$$\begin{aligned} &m_i v_{in}(t) + m_j v_{jn}(t) + \int_t^{t+\delta t} \boldsymbol{\mathcal{E}}_i \cdot \boldsymbol{n}_{ij}\, dt + \int_t^{t+\delta t} \boldsymbol{\mathcal{E}}_j \cdot \boldsymbol{n}_{ij}\, dt \\ &\quad = m_i v_{in}(t + \delta t) + m_j v_{jn}(t + \delta t), \end{aligned} \tag{3.5}$$

where the subscript n denotes the normal component of the velocity (along the line connecting particle centers) and the $\boldsymbol{\mathcal{E}}$'s represent all forces induced by near-field interaction with other particles, as well as all other external forces, if any, applied to the pair. If one isolates one of the members of the colliding pair, then

[1] Such forces can arise from surrounding fluid (treated in Zohdi [106]) or, as we later discuss, external electromagnetic fields where $\boldsymbol{\Psi}_i^{env} = q_i(\boldsymbol{E}^{ext} + \boldsymbol{v}_i \times \boldsymbol{B}^{ext})$.

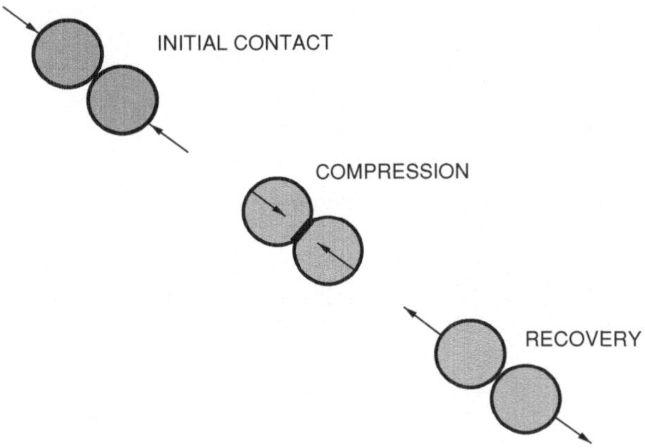

Fig. 3.1 Stages of impact: initial contact, compression and recovery of two impacting particles (Zohdi [110])

$$m_i v_{in}(t) + \int_t^{t+\delta t} I_n \, dt + \int_t^{t+\delta t} \mathcal{E}_i \cdot \boldsymbol{n}_{ij} \, dt = m_i v_{in}(t + \delta t), \tag{3.6}$$

where $\int_t^{t+\delta t} I_n \, dt$ is the total normal impulse due to impact. For a pair of particles undergoing impact, let us consider a decomposition of the collision event (Fig. 3.1) into a compression (δt_1) and recovery (δt_2) phase, i.e., $\delta t = \delta t_1 + \delta t_2$. Between the compression and recovery phases, the particles achieve a common normal velocity,[2] denoted v_{cn}, at the intermediate time $t + \delta t_1$. We may write for particle i, along the normal, in the compression phase of impact[3]

$$m_i v_{in}(t) + \int_t^{t+\delta t_1} I_n \, dt + \int_t^{t+\delta t_1} \mathcal{E}_i \cdot \boldsymbol{n}_{ij} \, dt = m_i v_{cn}, \tag{3.7}$$

and in the recovery phase

$$m_i v_{cn} + \int_{t+\delta t_1}^{t+\delta t} I_n \, dt + \int_{t+\delta t_1}^{t+\delta t} \mathcal{E}_i \cdot \boldsymbol{n}_{ij} \, dt = m_i v_{in}(t + \delta t). \tag{3.8}$$

For the other particle (j), in the compression phase,

$$m_j v_{jn}(t) - \int_t^{t+\delta t_1} I_n \, dt + \int_t^{t+\delta t_1} \mathcal{E}_j \cdot \boldsymbol{n}_{ij} \, dt = m_j v_{cn}, \tag{3.9}$$

---

[2] A common normal velocity for particles should be interpreted as indicating that the relative velocity in the normal direction between particle centers is zero.

[3] We assume that the "instantaneous" impact events do not occur at the same instant and that their action can be additively collected together within a time step.

and in the recovery phase

$$m_j v_{cn} - \int_{t+\delta t_1}^{t+\delta t} I_n\, dt + \int_{t+\delta t_1}^{t+\delta t} \boldsymbol{\mathcal{E}}_j \cdot \boldsymbol{n}_{ij}\, dt = m_j v_{jn}(t+\delta t). \tag{3.10}$$

This leads to an expression for the coefficient of restitution

$$\begin{aligned} e \overset{\text{def}}{=} \frac{\int_{t+\delta t_1}^{t+\delta t} I_n\, dt}{\int_t^{t+\delta t_1} I_n\, dt} &= \frac{m_i(v_{in}(t+\delta t) - v_{cn}) - \mathcal{E}_{in}(t+\delta t_1, t+\delta t)}{m_i(v_{cn} - v_{in}(t)) - \mathcal{E}_{in}(t, t+\delta t_1)} \\ &= \frac{-m_j(v_{jn}(t+\delta t) - v_{cn}) + \mathcal{E}_{jn}(t+\delta t_1, t+\delta t)}{-m_j(v_{cn} - v_{jn}(t)) + \mathcal{E}_{jn}(t, t+\delta t_1)}, \end{aligned} \tag{3.11}$$

where

$$\begin{aligned} \mathcal{E}_{in}(t+\delta t_1, t+\delta t) &\overset{\text{def}}{=} \int_{t+\delta t_1}^{t+\delta t} \boldsymbol{\mathcal{E}}_i \cdot \boldsymbol{n}_{ij}\, dt, \\ \mathcal{E}_{jn}(t+\delta t_1, t+\delta t) &\overset{\text{def}}{=} \int_{t+\delta t_1}^{t+\delta t} \boldsymbol{\mathcal{E}}_j \cdot \boldsymbol{n}_{ij}\, dt, \\ \mathcal{E}_{in}(t, t+\delta t_1) &\overset{\text{def}}{=} \int_{t}^{t+\delta t_1} \boldsymbol{\mathcal{E}}_i \cdot \boldsymbol{n}_{ij}\, dt, \\ \mathcal{E}_{jn}(t, t+\delta t_1) &\overset{\text{def}}{=} \int_{t}^{t+\delta t_1} \boldsymbol{\mathcal{E}}_j \cdot \boldsymbol{n}_{ij}\, dt. \end{aligned} \tag{3.12}$$

If we eliminate v_{cn}, we obtain an expression for e, that is,

$$e = \frac{v_{jn}(t+\delta t) - v_{in}(t+\delta t) + \Delta_{ij}(t+\delta t_1, t+\delta t)}{v_{in}(t) - v_{jn}(t) + \Delta_{ij}(t, t+\delta t_1)}, \tag{3.13}$$

where[4]

$$\Delta_{ij}(t+\delta t_1, t+\delta t) \overset{\text{def}}{=} \frac{1}{m_i}\mathcal{E}_{in}(t+\delta t_1, t+\delta t) - \frac{1}{m_j}\mathcal{E}_{jn}(t+\delta t_1, t+\delta t) \tag{3.15}$$

and

$$\Delta_{ij}(t, t+\delta t_1) \overset{\text{def}}{=} \frac{1}{m_i}\mathcal{E}_{in}(t, t+\delta t_1) - \frac{1}{m_j}\mathcal{E}_{jn}(t, t+\delta t_1). \tag{3.16}$$

---

[4] This collapses to the classical expression for the ratio of the relative velocities before and after impact if the near-field forces are negligible:

$$e \overset{\text{def}}{=} \frac{v_{jn}(t+\delta t) - v_{in}(t+\delta t)}{v_{in}(t) - v_{jn}(t)}. \tag{3.14}$$

Thus, we may rewrite Eq. 3.13 as

$$v_{jn}(t+\delta t) = v_{in}(t+\delta t) - \Delta_{ij}(t+\delta t_1, t+\delta t) + e\left(v_{in}(t) - v_{jn}(t) + \Delta_{ij}(t, t+\delta t_1)\right). \tag{3.17}$$

It is convenient to denote the average force (over time δt) acting on the particle from external sources as

$$\overline{\mathcal{E}}_{in} \stackrel{\text{def}}{=} \frac{1}{\delta t}\int_t^{t+\delta t} \boldsymbol{\mathcal{E}}_i \cdot \boldsymbol{n}_{ij}\, dt. \tag{3.18}$$

If e is explicitly known, then, combining Eqs. 3.13 and 3.5, one can write

$$v_{in}(t+\delta t) = \frac{m_i v_{in}(t) + m_j(v_{jn}(t) - e(v_{in}(t) - v_{jn}(t)))}{m_i + m_j} + \frac{(\overline{\mathcal{E}}_{in} + \overline{\mathcal{E}}_{jn})\delta t - m_j(e\Delta_{ij}(t, t+\delta t_1) - \Delta_{ij}(t+\delta t_1, t+\delta t))}{m_i + m_j}, \tag{3.19}$$

and, once $\boldsymbol{v}_{in}(t+\delta t)$ is known, one can subsequently also solve for $v_{jn}(t+\delta t)$ via Eq. 3.17.

Remark 1 Later, it will be useful to define the average impulsive normal contact force between the particles acting during the impact event as (using Eq. 3.6)

$$\overline{I}_n \stackrel{\text{def}}{=} \frac{1}{\delta t}\int_t^{t+\delta t} I_n\, dt = \frac{m_i(v_{in}(t+\delta t) - v_{in}(t))}{\delta t} - \overline{\mathcal{E}}_{in}. \tag{3.20}$$

In particular, as will be done later in the analysis, when we discretize the equations of motion with a discrete (finite difference) time step of Δt, where $\delta t \ll \Delta t$, we shall define the impulsive normal contact contribution to the total force acting on a particle, $\boldsymbol{\Psi}_i^{tot} = \boldsymbol{\Psi}_i^{nf} + \boldsymbol{\Psi}_i^{con} + \boldsymbol{\Psi}_i^{fric} + \boldsymbol{\Psi}_i^{env}$ (Eq. 3.1), to be[5]

$$\boldsymbol{\Psi}^{con} = \frac{\overline{I}_n \delta t}{\Delta t}\boldsymbol{n}_{ij}. \tag{3.21}$$

Furthermore, at the implementation level, we choose $\delta t = \gamma\Delta t$, where $0 < \gamma \ll 1$ and Δt is the time step discretization size, which will be introduced later in the work. A typical choice is $0 < \gamma \leq 0.01$. Typically, the system is insensitive to γ below 0.01. We assume $\delta t = \delta t_1 + \delta t_2 = \delta t_1 + e\delta t_1$, which immediately allows the following definitions, that is,

$$\delta t_1 = \frac{\gamma\Delta t}{1+e} \quad \text{and} \quad \delta t_2 = \frac{e\gamma\Delta t}{1+e}. \tag{3.22}$$

[5] One can interpret this as "smearing" out the force over the entire time step in which the collision occurs.

These results are consistent with the fact that the recovery time vanishes (all compression and no recovery) for $e \to 0$ (asymptotically "plastic") and, as $e \to 1$, the recovery time equals the compression time ($\delta t_2 = \delta t_1$, asymptotically "elastic"). If $e = 1$, there is no loss in energy, while if $e = 0$, there is a maximum loss in energy. For a more detailed analysis of impact duration times, see Johnson [33].

Remark 2 It is obvious that for a deeper understanding of the deformation within a particle, it must be treated as a deformable continuum, which will require a spatial discretization, for example, using the finite element method of the body (particle). For general references on the subject, see the well-known books of Bathe [7], Becker, Carey and Oden [8], Hughes [30], Szabo and Babuska [84] and Zienkiewicz and Taylor [94]. For a novel approach to finite element methods, see the recent work of Wriggers [91]. For work specifically focusing on the continuum mechanics of particles, see Zohdi and Wriggers [113]. For a detailed numerical analysis of multifield contact between bodies, see Wriggers [90].

3.2 "Friction" (Resistance to Sliding)

To incorporate frictional[6] stick-slip phenomena during impact for a general particle pair (i and j), the tangential velocities at the beginning of the impact time interval (time$=t$) are computed by subtracting the relative normal velocity away from the total relative velocity:

$$\boldsymbol{v}_{jt}(t) - \boldsymbol{v}_{it}(t) = (\boldsymbol{v}_j(t) - \boldsymbol{v}_i(t)) - \big((\boldsymbol{v}_j(t) - \boldsymbol{v}_i(t)) \cdot \boldsymbol{n}_{ij}\big)\, \boldsymbol{n}_{ij}. \tag{3.23}$$

One then writes the equation for tangential momentum change during impact for the ith particle

$$m_i v_{it}(t) - \overline{I}_f \delta t + \overline{\mathcal{E}}_{it} \delta t = m_i v_{ct}, \tag{3.24}$$

where the time average friction contribution is

$$\overline{I}_f = \frac{1}{\delta t} \int_t^{t+\delta t} I_f \, dt, \tag{3.25}$$

where the sum of the (time average) contributions from all other particles in the tangential direction ($\boldsymbol{\tau}_{ij}$) are

$$\overline{\mathcal{E}}_{it} = \frac{1}{\delta t} \int_t^{t+\delta t} \boldsymbol{\mathcal{E}}_i \cdot \boldsymbol{\tau}_{ij} \, dt \tag{3.26}$$

[6] It is probably more accurate to refer to this as "resistance to sliding" at such small scales, however, for brevity, we refer to the effect as "friction."

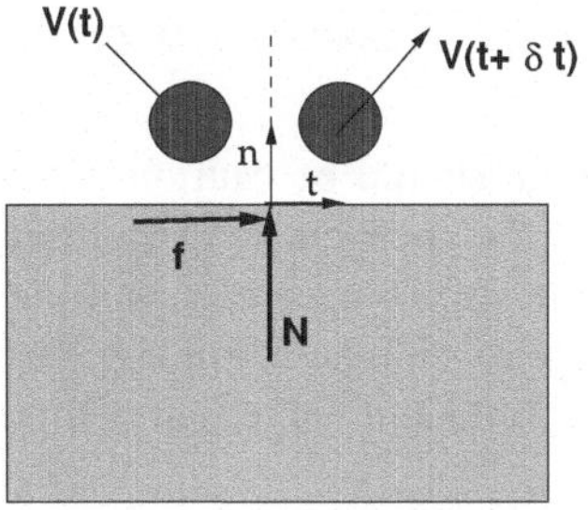

Fig. 3.2 A particle impacting a surface (Zohdi [110])

and where v_{ct} is the common velocity of particles i and j in the tangential direction.[7] Similarly, for the jth particle we have

$$m_j v_{jt}(t) + \overline{I}_f \delta t + \overline{\mathcal{E}}_{jt} \delta t = m_j v_{ct}. \tag{3.27}$$

There are two unknowns, $\overline{I}_f$ and v_{ct}. The main quantity of interest is $\overline{I}_f$, which can be solved for

$$\overline{I}_f = \frac{\left(\frac{\overline{\mathcal{E}}_{it}}{m_i} - \frac{\overline{\mathcal{E}}_{jt}}{m_j}\right) \delta t + v_{it}(t) - v_{jt}(t)}{\left(\frac{1}{m_i} + \frac{1}{m_j}\right) \delta t}. \tag{3.28}$$

Thus, consistent with stick-slip models of Coloumb friction, one first assumes no slip occurs. If

$$|\overline{I}_f| > \mu_s |\overline{I}_n|, \tag{3.29}$$

where the coefficient of static friction μ_s satisfies (μ_d is the dynamic coefficient of friction)

$$\mu_s \geq \mu_d, \tag{3.30}$$

then slip must occur and a dynamic sliding friction model is used. If sliding occurs, the friction force is assumed to be proportional to the normal force and opposite to the direction of relative tangent motion, i.e.,

$$\boldsymbol{\Psi}_i^{fric} \stackrel{\text{def}}{=} \mu_d ||\boldsymbol{\Psi}^{con}|| \frac{\boldsymbol{v}_{jt} - \boldsymbol{v}_{it}}{||\boldsymbol{v}_{jt} - \boldsymbol{v}_{it}||} = -\boldsymbol{\Psi}_j^{fric}. \tag{3.31}$$

Remark On the implementation level, as is done for the contact force, the friction force is "smeared" out over the entire time step in which the collision occurs ($\frac{\boldsymbol{\Psi}^{fric} \delta t}{\Delta t}$).

[7] They do not move relative to one another at this moment during the contact interval.

3.2.1 Restrictions on Friction Coefficients

One should be cautioned that when slipping occurs, there are limits on the values of the coefficients of friction for the problem to make sense. For example, consider the simple case of a particle (with radius small enough to neglect spin) (Fig. 3.2) approaching a stationary surface with velocity $\boldsymbol{v}(t)$, which can be decomposed into normal and tangential components,

$$\boldsymbol{v}(t) = v_n(t)\boldsymbol{e}_n + v_t(t)\boldsymbol{e}_t. \tag{3.32}$$

Now, consider the pre- and postimpact kinetic energy assuming sliding, (dynamic friction)

$$T(t) = \tfrac{1}{2}m(v_n^2(t) + v_t^2(t)), \tag{3.33}$$

and

$$T(t+\delta t) = \tfrac{1}{2}m(v_n^2(t+\delta t) + v_t^2(t+\delta t)). \tag{3.34}$$

Assuming sliding takes place, the impulse-momentum relation in the normal direction can be written as

$$mv_n(t) + \int_t^{t+\delta t} I_n\, dt = mv_n(t+\delta t), \tag{3.35}$$

and the tangential direction

$$mv_t(t) + \int_t^{t+\delta t} I_f\, dt = mv_t(t+\delta t). \tag{3.36}$$

For the normal direction (normal to the wall)

$$\int_t^{t+\delta t} I_n\, dt = m(v_n(t+\delta t) - v_n(t)) = -(1+e)mv_n(t) = \overline{I_n}\delta t, \tag{3.37}$$

where $\overline{I_n}$ is the average impulse over the impact time scale. For the tangential direction,

$$\int_t^{t+\delta t} I_f\, dt = m(v_t(t+\delta t) - v_t(t)) = \overline{I_f}\delta t. \tag{3.38}$$

Taking the limiting case of the largest possible coefficient of static friction, $I_f = -\mu_s I_n$, yields, with Eq. 3.37,

$$\overline{I_f} = \frac{m\mu_s(1+e)v_n(t)}{\delta t} = \frac{m}{\delta t}(v_t(t+\delta t) - v_t(t)), \tag{3.39}$$

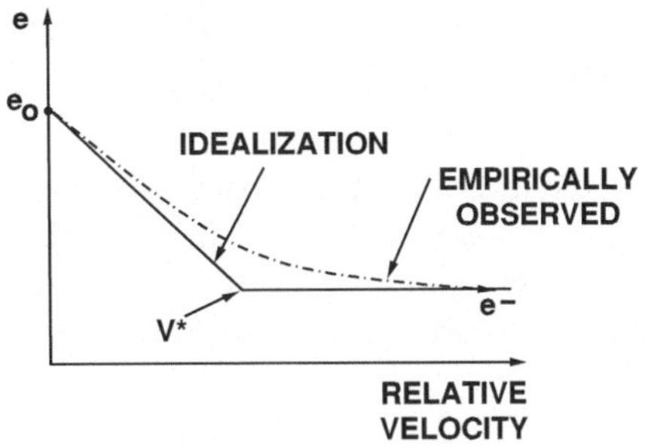

Fig. 3.3 Qualitative behavior of the coefficient of restitution with impact velocity (Zohdi [101])

which yields

$$v_t(t+\delta t) = v_t(t) + \mu_s(1+e)v_n(t). \tag{3.40}$$

Now, consider the restriction that the friction force cannot be so large that it reverses the initial tangential motion. Mathematically, this restriction can be written as

$$v_t(t+\delta t) = v_t(t) + \mu_s(1+e)v_n(t) \geq 0, \tag{3.41}$$

which leads to the following expression (note that $v_n(t) < 0$ and $v_t(t) > 0$)

$$\mu_s \leq -\frac{v_t(t)}{v_n(t)(1+e)}. \tag{3.42}$$

The same analysis can be repeated for the dynamic coefficient of friction, and not sliding backwards, yielding

$$\mu_d \leq \mu_s \leq -\frac{v_t(t)}{v_n(t)(1+e)}. \tag{3.43}$$

Thus, the coefficients of friction must be restricted in order to make physical sense. Qualitatively, as e grows, the restrictions on the coefficients of friction are more severe. For more general analyses on the validity of mechanical models involving friction see, for example, Oden and Pires [61], Martins and Oden [54], Kikuchi and Oden [39], Klarbring [41], Tuzun and Walton [62] or Cho and Barber [12].

Remark One can also determine the bound on the coefficient of friction by maximizing energy loss. $T(t+\delta t)$ can be minimized as a function of the friction coefficients when $v_t(t+\delta t) = 0$ because $v_n(t+\delta t)$ is independent of μ. This yields the same condition as before.

3.2.2 Velocity-Dependent Coefficients of Restitution

It is important to realize that, in reality, the phenomenological parameter e depends on the severity of the impact velocity. For extensive experimental data, see Goldsmith [28], or Johnson [33] for a more detailed analytical treatment. Qualitatively, the coefficient of restitution has behavior as shown in Fig. 3.3. A mathematical idealization of the behavior can be constructed as

$$e \stackrel{\text{def}}{=} max\left(e_o\left(1-\frac{\Delta v_n}{v^*}\right), e^-\right), \tag{3.44}$$

where v^* is a critical threshold velocity (normalization) parameter and where the relative velocity of approach is defined by

$$\Delta v_n \stackrel{\text{def}}{=} |v_{jn}(t) - v_{in}(t)| \tag{3.45}$$

and e^- is a lower limit to the coefficient of restitution.

3.3 Iterative Solution Schemes

3.3.1 General Time-Stepping Schemes

We now specifically address the second-order systems of interest. The equation of motion is given by

$$m_i \dot{\boldsymbol{v}}_i = \boldsymbol{\Psi}_i^{tot}, \tag{3.46}$$

where $\boldsymbol{\Psi}_i^{tot}$ is the total force provided from interactions with the external environment. Expanding the velocity in a Taylor series about $t + \phi\Delta t$, we obtain

$$\boldsymbol{v}_i(t+\Delta t) = \boldsymbol{v}_i(t+\phi\Delta t) + \frac{d\boldsymbol{v}_i}{dt}|_{t+\phi\Delta t}(1-\phi)\Delta t + \frac{1}{2}\frac{d^2\boldsymbol{v}_i}{dt^2}|_{t+\phi\Delta t}(1-\phi)^2(\Delta t)^2 + \mathcal{O}((\Delta t)^3) \tag{3.47}$$

and

$$\boldsymbol{v}_i(t) = \boldsymbol{v}_i(t+\phi\Delta t) - \frac{d\boldsymbol{v}_i}{dt}|_{t+\phi\Delta t}\phi\Delta t + \frac{1}{2}\frac{d^2\boldsymbol{v}_i}{dt^2}|_{t+\phi\Delta t}\phi^2(\Delta t)^2 + \mathcal{O}((\Delta t)^3). \tag{3.48}$$

Subtracting the two expressions yields

$$\frac{d\boldsymbol{v}_i}{dt}|_{t+\phi\Delta t} = \frac{\boldsymbol{v}_i(t+\Delta t) - \boldsymbol{v}_i(t)}{\Delta t} + \hat{\mathcal{O}}(\Delta t), \tag{3.49}$$

where $\hat{\mathcal{O}}(\Delta t) = \mathcal{O}((\Delta t)^2)$, when $\phi = \frac{1}{2}$. Thus, inserting this into the equations of equilibrium yields

$$\boldsymbol{v}_i(t+\Delta t) = \boldsymbol{v}_i(t) + \frac{\Delta t}{m_i}\boldsymbol{\Psi}^{tot}(t+\phi\Delta t) + \hat{\mathcal{O}}((\Delta t)^2). \tag{3.50}$$

Note that adding a weighted sum of Eqs. 3.47 and 3.48 yields

$$\boldsymbol{v}_i(t+\phi\Delta t) = \phi\boldsymbol{v}_i(t+\Delta t) + (1-\phi)\boldsymbol{v}_i(t) + \mathcal{O}((\Delta t)^2), \tag{3.51}$$

which will be useful shortly. Now, expanding the position of the center of mass in a Taylor series about $t+\phi\Delta t$, we obtain

$$\begin{aligned}\boldsymbol{r}_i(t+\Delta t) &= \boldsymbol{r}_i(t+\phi\Delta t) + \frac{d\boldsymbol{r}_i}{dt}|_{t+\phi\Delta t}(1-\phi)\Delta t \\ &\quad + \frac{1}{2}\frac{d^2\boldsymbol{r}}{dt^2}|_{t+\phi\Delta t}(1-\phi)^2(\Delta t)^2 + \mathcal{O}((\Delta t)^3)\end{aligned} \tag{3.52}$$

and

$$\boldsymbol{r}_i(t) = \boldsymbol{r}_i(t+\phi\Delta t) - \frac{d\boldsymbol{r}_i}{dt}|_{t+\phi\Delta t}\phi\Delta t + \frac{1}{2}\frac{d^2\boldsymbol{r}_i}{dt^2}|_{t+\phi\Delta t}\phi^2(\Delta t)^2 + \mathcal{O}((\Delta t)^3). \tag{3.53}$$

Subtracting the two expressions yields

$$\frac{\boldsymbol{r}_i(t+\Delta t) - \boldsymbol{r}_i(t)}{\Delta t} = \boldsymbol{v}_i(t+\phi\Delta t) + \hat{\mathcal{O}}(\Delta t). \tag{3.54}$$

Inserting Eq. 3.51 yields

$$\boldsymbol{r}_i(t+\Delta t) = \boldsymbol{r}_i(t) + (\phi\boldsymbol{v}_i(t+\Delta t) + (1-\phi)\boldsymbol{v}_i(t))\,\Delta t + \hat{\mathcal{O}}((\Delta t)^2), \tag{3.55}$$

and thus using Eq. 3.50 yields

$$\boldsymbol{r}_i(t+\Delta t) = \boldsymbol{r}_i(t) + \boldsymbol{v}_i(t)\Delta t + \frac{\phi(\Delta t)^2}{m_i}\boldsymbol{\Psi}_i^{tot}(t+\phi\Delta t) + \hat{\mathcal{O}}((\Delta t)^2). \tag{3.56}$$

The term $\boldsymbol{\Psi}_i^{tot}(t+\phi\Delta t)$ can be handled in two main ways:

- $\boldsymbol{\Psi}_i^{tot}(t+\phi\Delta t) \approx \boldsymbol{\Psi}_i^{tot}(\phi\boldsymbol{r}_i(t+\Delta t) + (1-\phi)\boldsymbol{r}_i(t))$ or
- $\boldsymbol{\Psi}_i^{tot}(t+\phi\Delta t) \approx \phi\boldsymbol{\Psi}_i^{tot}(\boldsymbol{r}_i(t+\Delta t)) + (1-\phi)\boldsymbol{\Psi}_i^{tot}(\boldsymbol{r}_i(t))$.

The differences are quite minute between either of the above, thus, for brevity, we choose the latter. In summary, we have the following:

$$\boldsymbol{r}_i(t+\Delta t) = \boldsymbol{r}_i(t) + \boldsymbol{v}_i(t)\Delta t + \frac{\phi(\Delta t)^2}{m_i}\big(\phi\boldsymbol{\Psi}_i^{tot}(\boldsymbol{r}_i(t+\Delta t)) \\ + (1-\phi)\boldsymbol{\Psi}^{tot}(\boldsymbol{r}_i(t))\big) + \hat{\mathcal{O}}((\Delta t)^2), \tag{3.57}$$

where

- when $\phi = 1$, then this is the (implicit) Backward Euler scheme, which is very stable (very dissipative) and $\hat{\mathcal{O}}((\Delta t)^2) = \mathcal{O}((\Delta t)^2)$ locally in time,
- when $\phi = 0$, then this is the (explicit) Forward Euler scheme, which is conditionally stable and $\hat{\mathcal{O}}((\Delta t)^2) = \mathcal{O}((\Delta t)^2)$ locally in time and
- when $\phi = 0.5$, then this is the (implicit) "Midpoint" scheme, which is stable and $\hat{\mathcal{O}}((\Delta t)^2) = \mathcal{O}((\Delta t)^3)$ locally in time.

3.4 Modification of the Time-Stepping Scheme for Impact

Consider

$$m_i\dot{\boldsymbol{v}}_i = \boldsymbol{\Psi}_i^{tot} = \boldsymbol{\Psi}_i^{nf} + \boldsymbol{\Psi}_i^{con} + \boldsymbol{\Psi}_i^{fric} + \boldsymbol{\Psi}_i^{env}. \tag{3.58}$$

Separating the impulsive and continuous forces and applying the *ϕ-method* leads to

$$\frac{\boldsymbol{v}_i(t+\Delta t) - \boldsymbol{v}_i(t)}{\Delta t} = \dot{\boldsymbol{v}}_i(t+\phi\Delta t) \tag{3.59}$$

and

$$\begin{aligned}\boldsymbol{v}_i(t+\Delta t) &= \boldsymbol{v}_i(t) + \frac{1}{m_i}\int_t^{t+\Delta t}\boldsymbol{\Psi}_i^{tot}\,dt \\ &= \boldsymbol{v}_i(t) + \frac{1}{m_i}\bigg(\int_t^{t+\Delta t}(\boldsymbol{\Psi}_i^{nf} + \boldsymbol{\Psi}_i^{env})\,dt + \int_t^{t+\delta t}\boldsymbol{\Psi}_i^{con}\,dt \\ &\qquad + \int_t^{t+\delta t}\boldsymbol{\Psi}_i^{fric}\,dt\bigg) \\ &\approx \boldsymbol{v}_i(t) + \frac{\Delta t}{m_i}\bigg(\phi(\boldsymbol{\Psi}_i^{nf}(t+\Delta t) + \boldsymbol{\Psi}_i^{env}(t+\Delta t)) \\ &\qquad + (1-\phi)(\boldsymbol{\Psi}_i^{nf}(t) + \boldsymbol{\Psi}_i^{env}(t))\bigg) \\ &\quad + \frac{\delta t}{m_i}\left(\overline{\boldsymbol{\Psi}}_i^{con}(t^*) + \overline{\boldsymbol{\Psi}}_i^{fric}(t^*)\right),\end{aligned} \tag{3.60}$$

where $t \leq t^* \leq t + \Delta t$. The position can be computed via

$$\frac{\boldsymbol{r}_i(t+\Delta t) - \boldsymbol{r}_i(t)}{\Delta t} \approx \boldsymbol{v}_i(t+\phi\Delta t) \Rightarrow \boldsymbol{r}_i(t+\Delta t) = \boldsymbol{r}_i(t) + \boldsymbol{v}_i(t+\phi\Delta t)\Delta t \tag{3.61}$$

and

$$\begin{aligned}\boldsymbol{r}_i(t+\Delta t) &= \boldsymbol{r}_i(t) + \boldsymbol{v}_i(t+\phi\Delta t)\Delta t \\ &= \boldsymbol{r}_i(t) + (\phi\boldsymbol{v}_i(t+\Delta t) + (1-\phi)\boldsymbol{v}_i(t))\,\Delta t.\end{aligned} \tag{3.62}$$

The position can be written as

$$\begin{aligned}\boldsymbol{r}_i(t+\Delta t) = \; & \boldsymbol{r}_i(t) + \boldsymbol{v}_i(t)\Delta t \\ & + \frac{\phi\Delta t}{m_i}\Big(\Big(\phi(\boldsymbol{\Psi}_i^{nf}(t+\Delta t) + \boldsymbol{\Psi}_i^{env}(t+\Delta t)) \\ & + (1-\phi)(\boldsymbol{\Psi}_i^{nf}(t) + \boldsymbol{\Psi}_i^{env}(t))\Big)\,\Delta t\Big) \\ & + \frac{\phi\Delta t}{m_i}\Big(\overline{\boldsymbol{\Psi}}_i^{con}(t^*)\delta t + \overline{\boldsymbol{\Psi}}_i^{fric}(t^*)\delta t\Big).\end{aligned} \tag{3.63}$$

Remark Generally, it is advantageous to follow a "collide and stream" philosophy, similar to lattice Boltzmann methods,[8] whereby collisions are evaluated at the end or beginning of the time step (i.e., $t^* = t$, updated at the end of the previous time step). This also mitigates large numbers of iterations within time steps if implicit schemes ($0 < \phi$) are used. Generally speaking, if a recursive process is *not employed* (an explicit scheme), the iterative error can accumulate rapidly. However, an overkill approach involving very small time steps, smaller than needed to control the discretization error, simply to suppress a nonrecursive process error, is computationally inefficient, and is discussed in detail in the next subsection.

3.4.1 Iterative (Implicit) Solution Methods

We write Eq. 3.63 in a slightly more streamlined form for particle i, that is,

$$\begin{aligned}\boldsymbol{r}_i^{L+1} = \; & \boldsymbol{r}_i^L + \boldsymbol{v}_i^L\Delta t \\ & + \frac{\phi\Delta t}{m_i}\Big(\Big(\phi(\boldsymbol{\Psi}_i^{nf,L+1} + \boldsymbol{\Psi}_i^{env,L+1}) + (1-\phi)(\boldsymbol{\Psi}^{nf,L} + \boldsymbol{\Psi}^{env,L})\Big)\,\Delta t\Big) \\ & + \frac{\phi\Delta t}{m_i}\Big(\overline{\boldsymbol{\Psi}}_i^{con}(t^*)\delta t + \overline{\boldsymbol{\Psi}}_i^{fric}(t^*)\delta t\Big),\end{aligned} \tag{3.64}$$

[8] See, for example, Sukop and Thorne [83] for a basic introduction to lattice Boltzmann methods.

which leads to a coupled set equations for $i = 1, 2, ..., N$ particles, where the superscript L is a time interval counter. The set of equations represented by Eq. 3.64 can be solved recursively.

3.4.2 Recursive Solution

Equation 3.64 can be solved recursively by recasting the relation as

$$\boldsymbol{r}_i^{L+1,K} = \mathcal{G}(\boldsymbol{r}_i^{L+1,K-1}) + \mathcal{R}_i, \tag{3.65}$$

where $K = 1, 2, 3, ...$ is the index of iteration within time step $L + 1$ and $\mathcal{R}_i$ is a remainder term that does not depend on the $(K + 1)$th solution, i.e.,

$$\mathcal{R}_i \neq \mathcal{R}_i(\boldsymbol{r}_1^{L+1}, \boldsymbol{r}_2^{L+1} ... \boldsymbol{r}_N^{L+1}). \tag{3.66}$$

The convergence of such a scheme is dependent on the behavior of $\mathcal{G}$. Namely, a sufficient condition for convergence is that $\mathcal{G}$ is a contraction mapping for all $\boldsymbol{r}_i^{L+1,K}$, $K = 1, 2, 3...$ In order to investigate this further, we define the iteration error as

$$(\mathbf{ERROR})_i^{L+1,K} \stackrel{\text{def}}{=} \boldsymbol{r}_i^{L+1,K} - \boldsymbol{r}_i^{L+1}. \tag{3.67}$$

A necessary restriction for convergence is iterative self consistency, i.e., the "exact" (discretized) solution must be represented by the scheme

$$\mathcal{G}(\boldsymbol{r}_i^{L+1}) + \mathcal{R}_i = \boldsymbol{r}_i^{L+1}. \tag{3.68}$$

Enforcing this restriction, a sufficient condition for convergence is the existence of a contraction mapping

$$\begin{aligned} \underbrace{||\boldsymbol{r}_i^{L+1,K} - \boldsymbol{r}_i^{L+1}||}_{(\mathbf{ERROR})_i^{L+1,K}} &= ||\mathcal{G}(\boldsymbol{r}_i^{L+1,K-1}) - \mathcal{G}(\boldsymbol{r}_i^{L+1})|| \\ &\leq \eta^{L+1,K} ||\boldsymbol{r}_i^{L+1,K-1} - \boldsymbol{r}_i^{L+1}||, \end{aligned} \tag{3.69}$$

where, if $0 \leq \eta^{L+1,K} < 1$ for each iteration K, then $(\mathbf{ERROR})_i^{L+1,K} \to \mathbf{0}$ for any arbitrary starting value $\boldsymbol{r}_i^{L+1,K=0}$, as $K \to \infty$. This type of contraction condition is sufficient, but not necessary, for convergence. Written out, the recursion is

$$\boldsymbol{r}_i^{L+1,K} = \underbrace{\boldsymbol{r}_i^L + \boldsymbol{v}_i^L \Delta t + \frac{\phi(\Delta t)^2}{m_i}\Big((1-\phi)(\boldsymbol{\Psi}_i^{nf}(\boldsymbol{r}^L) + \boldsymbol{\Psi}_i^{env}(\boldsymbol{r}^L))\Big)}_{\mathcal{R}}$$

$$+\underbrace{\frac{\phi(\Delta t)^2}{m_i}\left(\left(\phi(\boldsymbol{\Psi}_i^{nf,L+1,K-1}+\boldsymbol{\Psi}_i^{env,L+1,K-1})\right)\Delta t+\overline{\boldsymbol{\Psi}}_i^{con,K-1}(t^*)\delta t+\overline{\boldsymbol{\Psi}}_i^{fric,K-1}(t^*)\delta t\right)}_{\mathcal{G}(\boldsymbol{r}^{L+1,K-1})}, \tag{3.70}$$

where

$$\boldsymbol{\Psi}_i^{nf\ or\ env,L} \stackrel{\text{def}}{=} \boldsymbol{\Psi}_i^{nf\ or\ env,L}(\boldsymbol{r}_1^L,\boldsymbol{r}_2^L...\boldsymbol{r}_N^L) \tag{3.71}$$

and

$$\boldsymbol{\Psi}_i^{nf\ or\ env,L+1,K-1} \stackrel{\text{def}}{=} \boldsymbol{\Psi}_i^{nf\ or\ env,L+1,K-1}(\boldsymbol{r}_1^{L+1,K-1},\boldsymbol{r}_2^{L+1,K-1}...\boldsymbol{r}_N^{L+1,K-1}). \tag{3.72}$$

The convergence of Eq. 3.70 is scaled by

$$\eta \propto \frac{(\phi\Delta t)^2}{m_i}. \tag{3.73}$$

Therefore, we see that the contraction constant of $\mathcal{G}$ is (1) directly dependent on the strength of the interaction forces ($||\boldsymbol{\Psi}||$), (2) inversely proportional to m and (3) directly proportional to $(\Delta t)^2$ (at time$=t$). Therefore, if convergence is slow within a time step, the time step size, which is adjustable, can be reduced by an appropriate amount to increase the rate of convergence. Thus, decreasing the time step size improves the convergence, however, we want to simultaneously maximize the time step sizes to decrease overall computing time, while still meeting an error tolerance on the numerical solution's accuracy. In order to achieve this goal, we follow an approach found in Zohdi [95–113], originally developed for continuum thermochemical multifield problems, in which (1) one approximates

$$\eta^{L+1,K} \approx S(\Delta t)^p \tag{3.74}$$

(S is a constant) and (2) one assumes that the error within an iteration to behave according to

$$(S(\Delta t)^p)^K(\text{ERROR})^{L+1,0} = (\text{ERROR})^{L+1,K}, \tag{3.75}$$

$K = 1, 2, ...,$ where $(\text{ERROR})^{L+1,0} = ||\boldsymbol{r}^{L+1,K=1} - \boldsymbol{r}^L||$ is the initial norm of the iterative (relative) error and S is intrinsic to the system.[9] Our goal is to meet an error tolerance in exactly a preset (the analyst sets this) number of iterations. To this end, one writes

$$(S(\Delta t_{\text{tol}})^p)^{K_d}(\text{ERROR})^{L+1,0} = TOL, \tag{3.76}$$

[9] For the class of problems under consideration, due to the quadratic dependency on Δt, $p \approx 2$.

where TOL is a tolerance and where K_d is the number of desired iterations.[10] If the error tolerance is not met in the desired number of iterations, the contraction constant $\eta^{L+1,K}$ is too large. Accordingly, one can solve for a new smaller step size, under the assumption that S is constant,

$$\Delta t_{\text{tol}} = \Delta t \left(\frac{\left(\frac{TOL}{(\text{ERROR})^{L+1,0}}\right)^{\frac{1}{pK_d}}}{\left(\frac{(\text{ERROR})^{L+1,K}}{(\text{ERROR})^{L+1,0}}\right)^{\frac{1}{pK}}} \right). \tag{3.77}$$

The assumption that S is constant is not critical since the time steps are to be recursively refined and unrefined throughout the simulation. Clearly, the expression in Eq. 3.77 can also be used for time step enlargement, if convergence is met in less than K_d iterations. An implementation of the procedure is as follows:

(1) GLOBAL FIXED – POINT ITERATION : (SET i = 1 AND K = 0) :
(2) IF $i > N_p$ THEN GO TO (4)
(3) IF $i \leq N_p$ THEN :
  (a) COMPUTE POSITION : $\boldsymbol{r}_i^{L+1,K}$
  (b) GO TO (2) FOR NEXT PARTICLE (i = i + 1)
(4) ERROR MEASURE :
  $(a)(\text{ERROR})_K \stackrel{\text{def}}{=} \dfrac{\sum_{i=1}^{N_p} \|\boldsymbol{r}_i^{L+1,K} - \boldsymbol{r}_i^{L+1,K-1}\|}{\sum_{i=1}^{N_p} \|\boldsymbol{r}_i^{L+1,K} - \boldsymbol{r}_i^{L}\|}$ (normalized)
  $(b) Z_K \stackrel{\text{def}}{=} \dfrac{(\text{ERROR})_K}{TOL}$
  $(c) \Phi_K \stackrel{\text{def}}{=} \left(\dfrac{\left(\frac{TOL}{(\text{ERROR})_0}\right)^{\frac{1}{pK_d}}}{\left(\frac{(\text{ERROR})_K}{(\text{ERROR})_0}\right)^{\frac{1}{pK}}} \right)$
(5) IF TOLERANCE NOT MET ($Z_K > 1$) AND $K < K_d$ REPEAT ITERATION (K = K + 1)
(6) IF TOLERANCE MET ($Z_K \leq 1$) AND $K < K_d$ THEN :
  (a) INCREMENT TIME : $t = t + \Delta t$
  (b) CONSTRUCT NEW TIME STEP : $\Delta t = \Phi_K \Delta t$,
  (c) SELECT MINIMUM : $\Delta t = MIN(\Delta t^{lim}, \Delta t)$ AND GO TO (1)
(7) IF TOLERANCE NOT MET ($Z_K > 1$) AND $K = K_d$ THEN :
  (a) CONSTRUCT NEW TIME STEP : $\Delta t = \Phi_K \Delta t$
  (b) RESTART AT TIME = t AND GO TO (1).

(3.78)

Time step size adaptivity is important since the system's dynamics and configuration can dramatically change over the course of time, possibly requiring quite different time step sizes to control the iterative error. However, to maintain the accuracy of the time-stepping scheme, one must respect an upper bound dictated by the discretization error, i.e., $\Delta t \leq \Delta t^{lim}$.

[10] Typically, K_d is chosen to be between 5 and 10 iterations.

3.5 Thermal Effects and Coupled Systems

In certain applications, in addition to the near-field and contact effects introduced thus far, thermal behavior is of interest, for example, when the particles impact one another vigorously. In many cases, the source of heat generated during impact in such flows can be traced to the reactivity of the particles. This affects the mechanics of impact, for example, due to thermal softening. For instance, the presence of a reactive substance (e.g., hydrogen gas) adsorbed onto the surface can be a source of additional significant heat generation, through thermochemical reactions activated by impact forces, which thermally softens the material, thus reducing the coefficient of restitution , which in turn strongly affects the mechanical impact event itself (Fig. 3.4).

To illustrate how one can incorporate thermal effects, a somewhat ad hoc approach, building on the relation in Eq. 3.44, is to construct a thermally dependent coefficient of restitution as follows, using a multiplicative decomposition,

$$e \stackrel{\text{def}}{=} \left(max \left(e_o \left(1 - \frac{\Delta v_n}{v^*}\right), e^- \right)\right) \left(max \left(\left(1 - \frac{\theta}{\theta^*}\right), 0 \right)\right), \tag{3.78}$$

where θ is the temperature and θ^* can be considered as a thermal softening temperature. In order to determine the thermal state of the particles, we shall decompose the heat generation and heat transfer processes into two stages: (a) stage one, describing an extremely short time interval when impact occurs, $\delta t \ll \Delta t$, which accounts for the effects of chemical reactions and energy release due to mechanical straining and (b) stage two, which accounts for the postimpact behavior involving heat transfer between particles.

3.5.1 An Energy Balance

Consistent with the particle-based philosophy, it is assumed that the temperature fields are uniform within the particles. We consider an energy balance, governing the interconversions of mechanical, thermal and chemical energy in a system, dictated by the first law of thermodynamics. Accordingly, we require the time rate of change of the sum of the kinetic energy ($\mathcal{K}$) and stored energy ($\mathcal{S}$) to be equal to the work rate (power, $\mathcal{P}$), the heat flux due to conduction ($\mathcal{Q}$) and the net heat supplied, in this case, from two sources: (a) chemical ($\mathcal{H}$) and (b) impact-generated dissipation ($\mathcal{D}$). This leads to

$$\frac{d}{dt}(\mathcal{K} + \mathcal{S}) = \mathcal{P} + \mathcal{H} + \mathcal{D} + \mathcal{Q}, \tag{3.79}$$

where the stored energy is comprised of a thermal part,

$$\mathcal{S} = mC\theta, \tag{3.80}$$

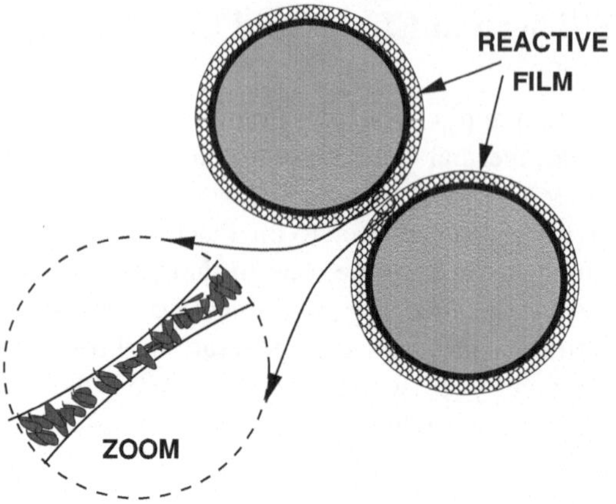

Fig. 3.4 Presence of dilute (smaller-scale) reactive gas particles adsorbed onto the surface of two impacting particles (Zohdi [103])

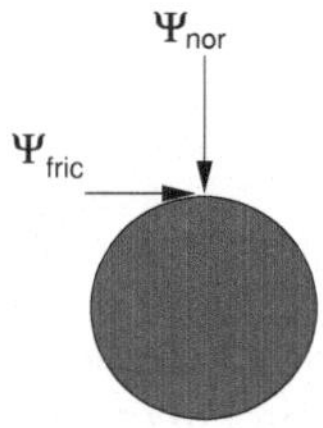

Fig. 3.5 Impulsive forces acting on a particle (Zohdi [110])

where C is the heat capacity per unit mass and, consistent with our assumptions that the particles deform negligibly during impact, we assume that there is an *insignificant amount of stored mechanical energy*. The kinetic energy is

$$\mathcal{K} = \frac{1}{2} m \boldsymbol{v} \cdot \boldsymbol{v}. \tag{3.81}$$

The mechanical power term is due to the forces acting on a particle, namely,

$$\mathcal{P} = \frac{d\mathcal{W}}{dt} = \boldsymbol{\Psi}^{tot} \cdot \boldsymbol{v}, \tag{3.82}$$

and, because

$$\frac{d\mathcal{K}}{dt} = m\dot{\boldsymbol{v}} \cdot \boldsymbol{v}, \tag{3.83}$$

and a balance of momentum

$$m\dot{\boldsymbol{v}} \cdot \boldsymbol{v} = \boldsymbol{\Psi}^{tot} \cdot \boldsymbol{v}, \tag{3.84}$$

we have

$$\frac{d\mathcal{K}}{dt} = \frac{d\mathcal{W}}{dt} = \mathcal{P}, \tag{3.85}$$

leading to

$$\frac{d\mathcal{S}}{dt} = \mathcal{H} + \mathcal{D} + \mathcal{Q}. \tag{3.86}$$

One can directly compute the work lost during the impact cycle (compression and recovery, Fig. 3.1), and hence the dissipation of mechanical energy to be (Fig. 3.5)

$$\int_t^{t+\delta t} \mathcal{D}_i \, dt \approx a_1 \overline{I}_n(t) v_n (1 - e^2)\delta t + a_2 \overline{I}_f(t) v_t (1 - e^2)\delta t \stackrel{\text{def}}{=} \overline{\mathcal{D}}_i \delta t, \tag{3.87}$$

where, if $e = 1$, there is no dissipation, e being the coefficient of restitution, and where $0 \leq a_1 \leq 1$ and $0 \leq a_2 \leq 1$ are dissipation parameters that indicate the proportion of mechanical dissipation that is converted to heat.

3.6 Numerical Scheme

We assume that the dynamics of any (dilute) gas does not affect the motion of the (much heavier) particles.[11] Also, we assume that radiative and convective terms are negligible. Thus, we only consider chemical, dissipative (due to impact) and conductive terms. Conduction is a continuous process, while the other two contributions are impulsive in nature since they owe their existence to impact. Thus,

$$m_i C_i \dot{\theta}_i = \mathcal{F}_i^{tot} \stackrel{\text{def}}{=} \mathcal{Q}_i + \mathcal{H}_i + \mathcal{D}_i, \tag{3.88}$$

where $\mathcal{Q}_i$ represents the conductive contribution from surrounding particles in contact. Separating the impulsive terms and continuous (conduction) term and integrating leads to, using

$$\dot{\theta}_i(t + \phi \Delta t) = \frac{\theta_i(t + \Delta t) - \theta_i(t)}{\Delta t}, \tag{3.89}$$

the following

$$\begin{aligned} \theta_i(t + \Delta t) &= \theta_i(t) + \frac{1}{m_i C_i} \int_t^{t+\Delta t} \mathcal{F}_i^{tot} \, dt \\ &= \theta_i(t) + \frac{1}{m_i C_i} \left(\int_t^{t+\Delta t} \mathcal{Q}_i \, dt + \int_t^{t+\delta t} \mathcal{H}_i \, dt + \int_t^{t+\delta t} \mathcal{D}_i \, dt \right) \\ &\approx \theta_i(t) + \frac{1}{m_i C_i} \left((\phi \mathcal{Q}_i(t + \Delta t) + (1 - \phi)\mathcal{Q}_i(t)) \Delta t + \overline{\mathcal{H}}_i(t^*)\delta t \right). \\ &\quad + \overline{\mathcal{D}}_i(t^*)\delta t \end{aligned} \tag{3.90}$$

[11] The gas only supplies a reactive thin film on the particles' surfaces.

3.6.1 Specific Relation for Reactions

The energy released from the reactions are assumed to be proportional to the amount of the surface substance available to be compressed in the contact area between the particles. A typical, ad hoc approximation in combustion processes is to write, for example,

$$\int_t^{t+\delta t} \mathcal{H}_i \, dt \approx \left(\kappa \min\left(\frac{|\overline{I}_n|}{I_n^*}, 1 \right) \pi b^2 \right) \delta t, \tag{3.91}$$

where $\overline{I}_n$ is the normal impact force, κ is a reaction (saturation) constant, energy per unit area, I_n^* is a normalization parameter and b is the particle radius. For details, for example, see Schmidt [80].[12] *Clearly, these equations are coupled to those of impact through the coefficient of restitution and the velocity-dependent impulse.* Additionally, the postcollision velocities are computed from the momentum relations which are coupled to the temperature. Later in the analysis, this equation is incorporated into an overall staggered fixed-point iteration scheme, whereby the temperature is predicted for a given velocity field, and then the velocities are recomputed with the new temperature field, and so on. The process is repeated until the fields change negligibly between successive iterations. The entire set of equations is embedded within a larger system, involving the particle dynamics, later in the analysis, and is solved in a recursive, staggered, manner.

3.6.2 Specific Relation for Conduction

It is assumed that the temperature fields are uniform within the (small) particles. We remark that the validity of using a lumped thermal model, i.e., ignoring temperature gradients and assuming a uniform temperature within a particle, is dictated by the magnitude of the Biot number. A small Biot number indicates that such an approximation is reasonable. The Biot number for spheres scales with the ratio of the particle volume (V) to the particle surface area (a_s), $\frac{V}{a_s} = \frac{b}{3}$, which indicates that a uniform temperature distribution is appropriate since the particles, by definition, are small. In continuous models of conduction, Fourier's law leads to

$$\mathcal{Q}_i = I\!K_i \nabla^2 \theta_i, \tag{3.92}$$

which can be interpreted as the value of the temperature of the node subtracted from the surrounding nodes' temperatures, where $I\!K_i$ is the effective particle conductivity.

[12] By construction, this model has increased heat production, via $\delta\mathcal{H}$, for increasing κ.

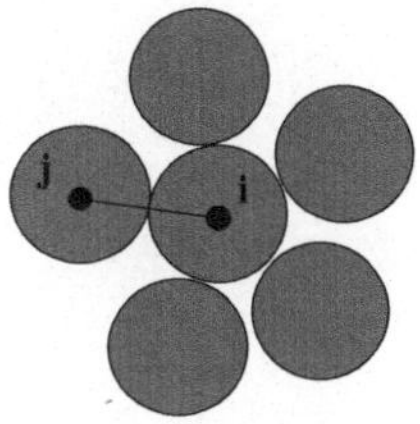

Fig. 3.6 Conduction between particles (Zohdi [110])

For example, for a three-dimensional finite-difference stencil commonly used in numerical methods, we have

$$\mathcal{Q}_i = I\!K_i \nabla^2 \theta_i \approx I\!K_i (\frac{\theta(x+\Delta x, y, z) - 2\theta(x, y, z) + \theta(x-\Delta x, y, z)}{(\Delta x)^2} + \frac{\theta(x, y+\Delta y, z) - 2\theta(x, y, z) + \theta(x, y-\Delta y, z)}{(\Delta y)^2} + \frac{\theta(x, y, z+\Delta z) - 2\theta(x, y, z) + \theta(x, y, z-\Delta z)}{(\Delta z)^2}), \tag{3.93}$$

which can be "generalized" as

$$\mathcal{Q}_i = I\!K_i \sum_{j=1}^{N_c} \frac{(\theta_j - \theta_i)}{||\boldsymbol{r}_j - \boldsymbol{r}_i||^2}, \tag{3.94}$$

where the summation extends over all particles $j = 1, 2, 3 ... N_c$ that are in contact with particle i (Fig. 3.6).

The time-stepping formula has the following form

$$\begin{aligned} \theta_i(t+\Delta t) = \; & \theta_i(t) + \frac{\delta t}{m_i C_i} \left(\overline{\mathcal{H}}_i + \overline{\mathcal{D}}_i \right) \\ & + \frac{\Delta t}{m_i C_i} \left(\phi I\!K_i \sum_{j=1}^{N_c} \frac{\theta_j(t+\Delta t) - \theta_i(t+\Delta t)}{||\boldsymbol{r}_j(t+\Delta t) - \boldsymbol{r}_i(t+\Delta t)||^2} \right) \\ & + \frac{\Delta t}{m_i C_i} \left((1-\phi) I\!K_i \sum_{j=1}^{N_c} \frac{\theta_j(t) - \theta_i(t)}{||\boldsymbol{r}_j(t) - \boldsymbol{r}_i(t)||^2} \right) \end{aligned} \tag{3.95}$$

which can be rewritten, isolating $\theta_i(t+\Delta t)$, as

$$\begin{aligned}\theta_i(t+\Delta t) = &\ A\theta_i(t) + \frac{A\delta t}{m_i C_i}\left(\overline{\mathcal{H}}_i + \overline{\mathcal{D}}_i\right) \\ &+ \frac{A\Delta t}{m_i C_i}\left(\phi \mathbb{K}_i \sum_{j=1}^{N_c} \frac{\theta_j(t+\Delta t)}{||\boldsymbol{r}_j(t+\Delta t) - \boldsymbol{r}_i(t+\Delta t)||^2}\right) \\ &+ \frac{A\Delta t}{m_i C_i}\left((1-\phi)\mathbb{K}_i \sum_{j=1}^{N_c} \frac{\theta_j(t) - \theta_i(t)}{||\boldsymbol{r}_j(t) - \boldsymbol{r}_i(t)||^2}\right)\end{aligned} \tag{3.96}$$

where

$$A = \left(1 + \frac{\phi \mathbb{K}_i \Delta t}{m_i C_i} \sum_{j=1}^{N_c} \frac{1}{||\boldsymbol{r}_j(t+\Delta t) - \boldsymbol{r}_i(t+\Delta t)||^2}\right)^{-1}. \tag{3.97}$$

An iterative scheme can be set up $K = 1, 2, 3..$,

$$\begin{aligned}\theta_i^{K+1}(t+\Delta t) = &\ A^K\theta_i(t) + \frac{A^K\delta t}{m_i C_i}\left(\overline{\mathcal{H}}_i^K + \overline{\mathcal{D}}_i^K\right) \\ &+ \frac{A^K\Delta t}{m_i C_i}\left(\phi \mathbb{K}_i \sum_{j=1}^{N_c} \frac{\theta_j^K(t+\Delta t)}{||\boldsymbol{r}_j^K(t+\Delta t) - \boldsymbol{r}_i^K(t+\Delta t)||^2}\right) \\ &+ \frac{A^K\Delta t}{m_i C_i}\left((1-\phi)\mathbb{K}_i \sum_{j=1}^{N_c} \frac{\theta_j(t) - \theta_i(t)}{||\boldsymbol{r}_j(t) - \boldsymbol{r}_i(t)||^2}\right).\end{aligned} \tag{3.98}$$

We note that Eq. 3.98 is of the general form:

$$\theta(t+\Delta t) = \mathcal{G}(\theta(t+\Delta t)) + \mathcal{R}, \tag{3.99}$$

where $\mathcal{R} \neq \mathcal{R}(\theta(t+\Delta t))$, and where $\mathcal{G}$'s behavior is controlled by the magnitude of Δt. Clearly, the temperature is coupled to the mechanical behavior of the system. Next, we develop a multiphysical staggering scheme to solve the overall system.

3.7 Multiphysical Staggering Scheme

Broadly speaking, staggering schemes proceed by solving each field equation individually, allowing only the primary field variable to be active. After the solution of each field equation, the primary field variable is updated, and the next field equation is addressed in a similar manner. Such approaches have a long history in the computational mechanics community. For example, see Park and Felippa [69], Zienkiewicz

[92], Zienkiewicz et al. [93], Schrefler [81], Turska and Schrefler [87], Lewis et al. [49], Doltsinis [15], [16], Piperno [71], Lewis and Schrefler [48], Armero and Simo [3–5], Armero [6], Le Tallec and Mouro [47], Zohdi [95–113] and the extensive works of Farhat and coworkers [72, 22, 46, 23, 73, 24]. For recent work involving staggering schemes for piezoelectric applications, see Fish and Chen [25]. Also, for a review of the most novel approaches, see Michopoulos et al. [56]. Generally speaking, if a recursive staggering process is *not employed* (an explicit coupling scheme), the staggering error can accumulate rapidly. However, an overkill approach (involving very small time steps, smaller than needed to control the discretization error), simply to suppress a nonrecursive staggering process error, is computationally inefficient. Therefore, the objective of the next subsection is to develop a strategy to adaptively adjust, in fact maximize, the choice of the time step size in order to control the staggering error, while simultaneously staying below a critical time step size needed to control the discretization error (as stated before). An important related issue is to simultaneously minimize the computational effort involved. The number of times the multifield system is solved, as opposed to time steps, is taken as the measure of computational effort since within a time step, many multifield system re-solves can take place. We now further develop the staggering scheme introduced earlier by extending an approach found in Zohdi [95–113] .

3.7.1 A General Iterative Framework

We consider an abstract setting, whereby one solves for the particle positions, assuming that the thermal fields are fixed,

$$\mathcal{A}_1(\underline{\boldsymbol{r}^{L+1,K}}, \theta^{L+1,K-1}) = \mathcal{F}_1(\boldsymbol{r}^{L+1,K-1}, \theta^{L+1,K-1}). \tag{3.100}$$

Then, one solves for the thermal fields, assuming the particle positions fixed,

$$\mathcal{A}_2(\boldsymbol{r}^{L+1,K}, \underline{\theta^{L+1,K}}) = \mathcal{F}_2(\boldsymbol{r}^{L+1,K}, \theta^{L+1,K-1}), \tag{3.101}$$

where only the underlined variable is "active," L indicates the time step and K indicates the iteration counter. Within the staggering scheme, implicit time-stepping methods, with time step size adaptivity, will be used throughout the upcoming analysis.

Continuing where Eq. 3.76 left off, we define the normalized errors within each time step, for the two fields,

$$(\text{ERROR})_{rK} \stackrel{\text{def}}{=} \frac{||\boldsymbol{r}^{L+1,K} - \boldsymbol{r}^{L+1,K-1}||}{||\boldsymbol{r}^{L+1,K} - \boldsymbol{r}^{L}||} \quad \text{and}$$
$$(\text{ERROR})_{\theta K} \stackrel{\text{def}}{=} \frac{||\theta^{L+1,K} - \theta^{L+1,K-1}||}{||\theta^{L+1,K} - \theta^{L}||}. \tag{3.102}$$

We define maximum "violation ratio," i.e., as the larger of the ratios of each field variable's error to its corresponding tolerance, by $Z_K \stackrel{\text{def}}{=} max(z_{rK}, z_{\theta K})$, where

$$z_{rK} \stackrel{\text{def}}{=} \frac{(\text{ERROR})_{rK}}{TOL_r} \quad \text{and} \quad z_{\theta K} \stackrel{\text{def}}{=} \frac{(\text{ERROR})_{\theta K}}{TOL_\theta}, \tag{3.103}$$

with the minimum scaling factor defined as $\Phi_K \stackrel{\text{def}}{=} min(\phi_{rK}, \phi_{\theta K})$, where

$$\phi_{rK} \stackrel{\text{def}}{=} \left(\frac{\left(\frac{TOL_r}{(\text{ERROR})_{r0}}\right)^{\frac{1}{pK_d}}}{\left(\frac{(\text{ERROR})_{rK}}{(\text{ERROR})_{r0}}\right)^{\frac{1}{pK}}} \right), \quad \phi_{\theta K} \stackrel{\text{def}}{=} \left(\frac{\left(\frac{TOL_\theta}{(\text{ERROR})_{\theta 0}}\right)^{\frac{1}{pK_d}}}{\left(\frac{(\text{ERROR})_{\theta K}}{(\text{ERROR})_{\theta 0}}\right)^{\frac{1}{pK}}} \right). \tag{3.104}$$

The algorithm is as follows:

(1) GLOBAL FIXED − POINT ITERATION : (SET i = 1 AND K = 0) :
(2) IF i > N_p THEN GO TO (4)
(3) IF i ≤ N_p THEN : (FOR PARTICLE i)
 (a) COMPUTE POSITION : $\boldsymbol{r}_i^{L+1,K}$
 (b) COMPUTE TEMPERATURE : $\theta_i^{L+1,K}$
 (c) GO TO (2) AND NEXT PARTICLE (i = i + 1)
(4) ERROR MEASURES(normalized) :

$$(a)(\text{ERROR})_{rK} \stackrel{\text{def}}{=} \frac{\sum_{i=1}^{N_p} ||\boldsymbol{r}_i^{L+1,K} - \boldsymbol{r}_i^{L+1,K-1}||}{\sum_{i=1}^{N_p} ||\boldsymbol{r}_i^{L+1,K} - \boldsymbol{r}_i^{L}||}$$

$$(\text{ERROR})_{\theta K} \stackrel{\text{def}}{=} \frac{\sum_{i=1}^{N_p} ||\theta_i^{L+1,K} - \theta_i^{L+1,K-1}||}{\sum_{i=1}^{N_p} ||\theta_i^{L+1,K} - \theta_i^{L}||}$$

$$(b) Z_K \stackrel{\text{def}}{=} max(z_{rK}, z_{\theta K}) \quad where \quad z_{rK} \stackrel{\text{def}}{=} \frac{(\text{ERROR})_{rK}}{TOL_r}, \quad z_{\theta K} \stackrel{\text{def}}{=} \frac{(\text{ERROR})_{\theta K}}{TOL_\theta}$$

$$(c) \Phi_K \stackrel{\text{def}}{=} min(\phi_{rK}, \phi_{\theta K}) \quad where$$

$$\phi_{rK} \stackrel{\text{def}}{=} \left(\frac{\left(\frac{TOL_r}{(\text{ERROR})_{r0}}\right)^{\frac{1}{pK_d}}}{\left(\frac{(\text{ERROR})_{rK}}{(\text{ERROR})_{r0}}\right)^{\frac{1}{pK}}} \right), \quad \phi_{\theta K} \stackrel{\text{def}}{=} \left(\frac{\left(\frac{TOL_\theta}{(\text{ERROR})_{\theta 0}}\right)^{\frac{1}{pK_d}}}{\left(\frac{(\text{ERROR})_{\theta K}}{(\text{ERROR})_{\theta 0}}\right)^{\frac{1}{pK}}} \right)$$

(5) IF TOLERANCE NOT MET ($Z_K > 1$) AND $K < K_d$ REPEAT ITERATION (K = K + 1)
(6) IF TOLERANCE MET ($Z_K \leq 1$) AND $K < K_d$ THEN :
 (a) INCREMENT TIME : $t = t + \Delta t$
 (b) CONSTRUCT NEW TIME STEP : $\Delta t = \Phi_K \Delta t$,
 (c) SELECT MINIMUM : $\Delta t = MIN(\Delta t^{lim}, \Delta t)$ AND GO TO (1)
(7) IF TOLERANCE NOT MET ($Z_K > 1$) AND $K = K_d$ THEN :
 (a) CONSTRUCT NEW TIME STEP : $\Delta t = \Phi_K \Delta t$
 (b) RESTART AT TIME = t AND GO TO (1).

(3.105)

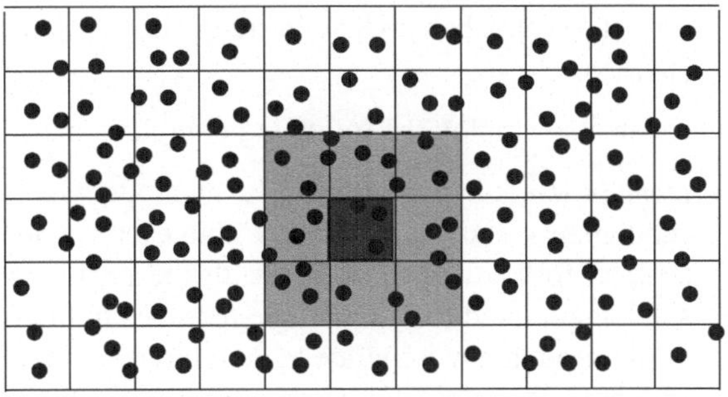

Fig. 3.7 Binning of the particles in a flow (Zohdi [110])

The overall goal is to deliver solutions where the staggering (incomplete coupling) error is controlled and the temporal discretization accuracy dictates the upper limits on the time step size (Δt^{lim}).

Remark A significant speed up in the computation can be achieved via Sorting and Binning (SB) methods. SB methods proceed by partitioning the whole domain into bins. The particles are sorted by the bins in which they reside. The particle interaction proceeds, bin by bin, where the particles within a bin only potentially interact with particles in other nearest neighbor bins. Essentially, for a given particle in a bin, contact searches and near-field interaction searches are conducted with particles in the neighboring bins only. The approach is relatively straightforward to implement and can speed up the computation dramatically. In order to determine the approximate savings, let us denote the number of particles by N_p. The computational costs are as follows:

- To determine (sort) which particle belong to which bins: $\mathcal{O}(N_p)$,
- For a bin, for each particle in the bin, search over only the particles in the neighboring bins. For example, consider an evenly distributed set of particles on Fig. 3.7. Denote the number of bins by N_b, and the number of nearest-neighbor bins by N_{nn}. The total cost for searching (to compute the particle interactions) and sweeping through all of the bins, is

$$\left(\frac{N_p}{N_b}\right)\left(N_{nn}\left(\frac{N_p}{N_b}\right)\right)N_b = N_p^2\left(\frac{N_{nn}}{N_b}\right). \tag{3.106}$$

  For example, for an *immediate nearest neighbor* list, $N_{nn} = 27$ (in 3-D, including the central bin).
  Note 1: One can reduce computation by accounting for particle-particle interaction that has already been computed from previous bin computation, $i - j$, when computing $j - i$ interaction.
  Note 2: One can assume that particles stay in the bins for a few time steps, and that one does not need to re-sort immediately. One can construct so-called "interaction" or "Verlet" lists of neighboring particles which particles interact with, for a few

time steps, and then update the interaction periodically (see Pöschel and Schwager [74]).

One can also employ the following complementary techniques:

- *Domain decomposition:* involve methods where the domain is partitioned into subdomains, and the particles within each subdomain are sent to a processor and stepped forward in time, but with the positions of the particles outside of the subdomain fixed (relative to the particles in that subdomain). This is done for all of the subdomains separately, then the position of all of the particles are updated and this information is shared between processors, then the process is repeated.
- *Fast (summation) multipole techniques:* involve methods that compute far-field interaction rapidly by exploiting the separable (near- and far-field) structure of multipole interaction.

Domain decomposition and fast multipole techniques were not employed here, though they are relatively easy to implement.

3.8 Model Problems

3.8.1 Initial Configurations: Preprocessing

In order to generate an initial particle configuration, one typically uses a classical random sequential addition algorithm of wisdom [89], placing nonoverlapping particles randomly into the domain of interest. However, the RSA algorithm usually cannot achieve extremely high density (volume fraction) configurations. For high-density sprays, the well-known, equilibrium-driven, Metropolis algorithm is widely used. However, for extremely high volume fractions, effectively packing (and "jamming") particles together, a relatively new class of methods, based on simultaneous particle flow and growth, has been developed by Torquato and coworkers [86, 36, 17–19], which are computationally efficient and straightforward to implement (Fig, 3.8).

The algorithm proceeds by allowing the particles to "grow" in the flow as the particles are moving, thus permitting the particles to slightly "bump" and rearrange themselves through momentum exchange. The growth rate is selected to be very gradual. For example, before a "plume" of particles is ready for an impact simulation, one can employ periodic boundary conditions (where the exiting particles are "re-fed" into the system), and at the end of each time step in this preliminary simulation, one can adjust the radius of the ith particle via

$$\boldsymbol{r}_i(t + \Delta t) = \boldsymbol{r}_i(t = 0)\left(1 + \mathcal{L}\frac{t}{T}\right), \tag{3.107}$$

where $\mathcal{L}$ is a user selected growth rate, and T is the total simulation "growth time." When the volume fraction has reached a sufficiently high level (this is determined

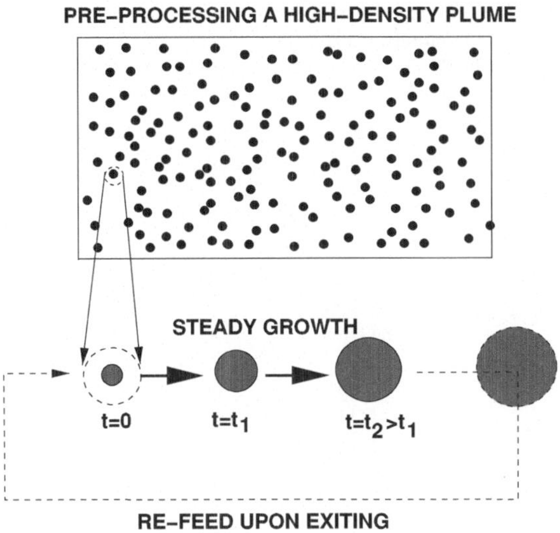

Fig. 3.8 Growth schemes (Zohdi [110])

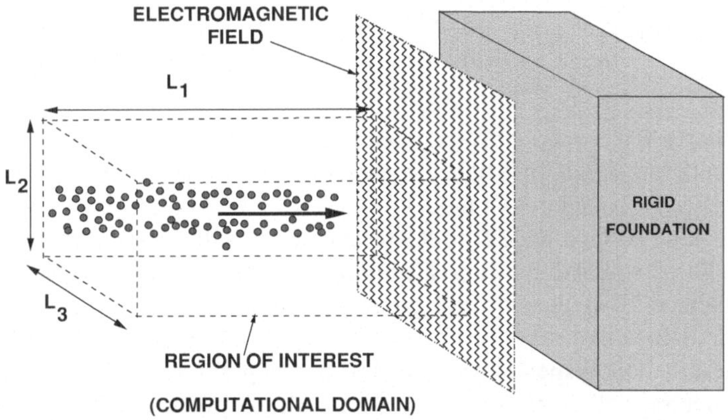

Fig. 3.9 A model problem: a particulate jet-spray impinging onto a (**1**) neutral surface, and (**2**) an electromagnetic "wall" (Zohdi [110])

by selecting $\mathcal{L}$ and T), the plume may then be used for the impact simulation with an external object.

3.8.2 Numerical Examples

We now reconsider the dynamics of particles in the presence of electromagnetic fields

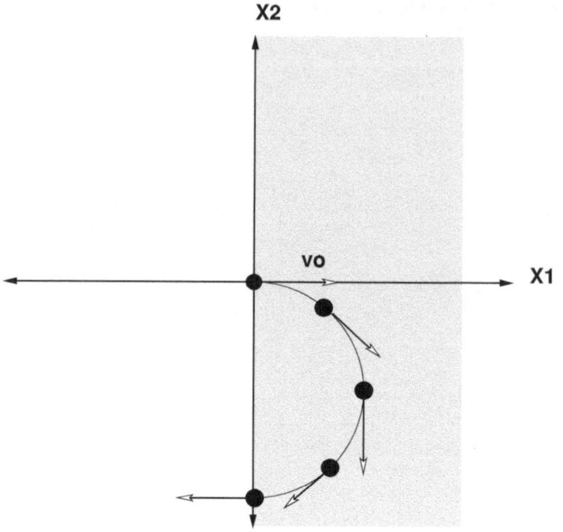

Fig. 3.10 Trajectory of the particle encountering a B-field (Zohdi [110])

$$m_i \ddot{\boldsymbol{r}}_i = \boldsymbol{\Psi}_i^{tot}(\boldsymbol{r}_1, \boldsymbol{r}_2, ..., \boldsymbol{r}_{N_p}) = \boldsymbol{\Psi}_i^{nf} + \boldsymbol{\Psi}_i^{con} + \boldsymbol{\Psi}_i^{fric} + \underbrace{q_i(\boldsymbol{E}^{ext} + \boldsymbol{v}_i \times \boldsymbol{B}^{ext})}_{\boldsymbol{\Psi}_i^{env}}, \tag{3.108}$$

where the fields $\boldsymbol{\Psi}_i^{env} = q_i(\boldsymbol{E}^{ext} + \boldsymbol{v}_i \times \boldsymbol{B}^{ext})$ are *externally controlled.* The presence of the Lorentz force can cause helical-type motion to occur, which will be discussed later. Specifically, we address a model problem of a particulate jet-spray impinging onto a (1) neutral surface, and (2) an electromagnetic "wall," placed before the surface (Fig. 3.9). The absolute dimensions are unimportant for the model problem, and have been normalized.[13] We considered a group of N_p randomly positioned particles in a jet-spray domain with dimensions $L_1 = 20 \times L_2 = 0.75 \times L_3 = 0.75$ m. The initial particle radius (monodisperse) was $r = 0.05$ m and preprocessed ("grown," without near-field neutral collisions) to $r = 0.1$ m, with the previously mentioned steady growth algorithm; which was then used for the simulations. The relevant simulation parameters were:

- number of particles = 1500, with $\alpha_{ij} = \bar{\alpha}_{ij} q_i q_j c_i c_j$ and $c_i = \pm 1$ (positive/negative), $q_i = q_j = 1$,
- $\bar{\alpha}_{ij1} = 2$, $\bar{\alpha}_{ij2} = 1$, $\beta_{ij1} = 1$, $\beta_{ij2} = 2$,
- mass density of the particles $= 2000$ kg/m^3,
- initial velocity = (30, 0, 0) m/s,
- initial mean position = (−5, 0, 0) m,
- coefficient of dynamic friction, $\mu_d = 0.1$,
- coefficient of static friction, $\mu_s = 0.2$,
- baseline coefficient of restitution, $e_o = 0.5$,

[13] The transverse dimensions of the jet were set to be approximately unity, initially. All system parameters can be scaled to describe any specific system of interest.

- limit of coefficient of restitution, $e^- = 0.2$,
- reaction constant, $\kappa = 10^4\,\mathrm{J/m^2}$,
- thermal constant, $\theta^* = 3000$ K,
- velocity parameter, $v^* = 10$ m/s,
- impact parameter, $I^* = 1000$ N,
- conductivity, $\mathbb{K}_i = 100\,\mathrm{Jm^2/s\,kg}$,
- $\theta(t = 0) = 300$,
- heat capacity, $C = 10^3$ J/kg K,
- impulse-thermal conversion, $a_1 = 1$, $a_2 = 1$,
- target number of fixed point iterations, $K_d = 6$,
- the time-stepping parameter, $\phi = 0.5$,
- $\boldsymbol{E}^{ext} = (0, 0, 0)$N/Coulomb,
- $\boldsymbol{B}^{ext} = (0, 0, 1000)$ kg/s Coulomb,
- domain bin-grid = (70,40,40),
- simulation duration = 1 s,
- initial time step size = 0.001 s,
- time step upper bound = 0.0025 s,
- tolerance for the fixed-point iteration = 10^{-3}.

Particles that strayed outside of a computational $4m \times 4m$ window were thrown out of the computations. We have the following observations:

- Figures 3.11, 3.12 and 3.13 illustrate the evolution of the microstructure of a jet-spray: a. interparticle collisions and heating b. agglomeration and c. impact with the surface or electromagnetic "wall."
- Depending on the time allowed for the jet-spray to impact the obstacle, there may not be enough time for clustering to fully evolve. We note that one can break up large detrimental clusters with magnetic fields by separating positive and negative charge clusters that have bonded together. In the case considered, the electromagnetic field is strong enough to repel the particulates. We remark that several different scenarios could be considered by varying the initial kinetic energy of the jet, for example, (1) epitaxy (particle lay-up) and (2) particle embedding/infiltration into the wall. Also, depending on the external fields, for example, in the vicinity of the surface, a pure E-field will draw in either positive or negative charged particles (thus separating them), although in this particular example, the E-field was set to zero.
- The key signature of the onset of clumping/clustering is the sudden spike in the impact force over time (Fig. 3.14), as opposed to the force "signature" produced by a steady stream of particles, which is more or less constant.
- The system achieves a much greater temperature (particle system averaged) in the case of the neutral wall (Fig. 3.19) due to the particle impact against the wall.
- One can compute the discrete "vortex" (angular momentum about the mass center) of the system:

$$\boldsymbol{V} \stackrel{\text{def}}{=} \frac{1}{\mathcal{M}} \sum_{i=1}^{N_p} m_i(\boldsymbol{r}_i - \boldsymbol{r}_{cm}) \times (\boldsymbol{v}_i - \boldsymbol{v}_{cm}), \tag{3.109}$$

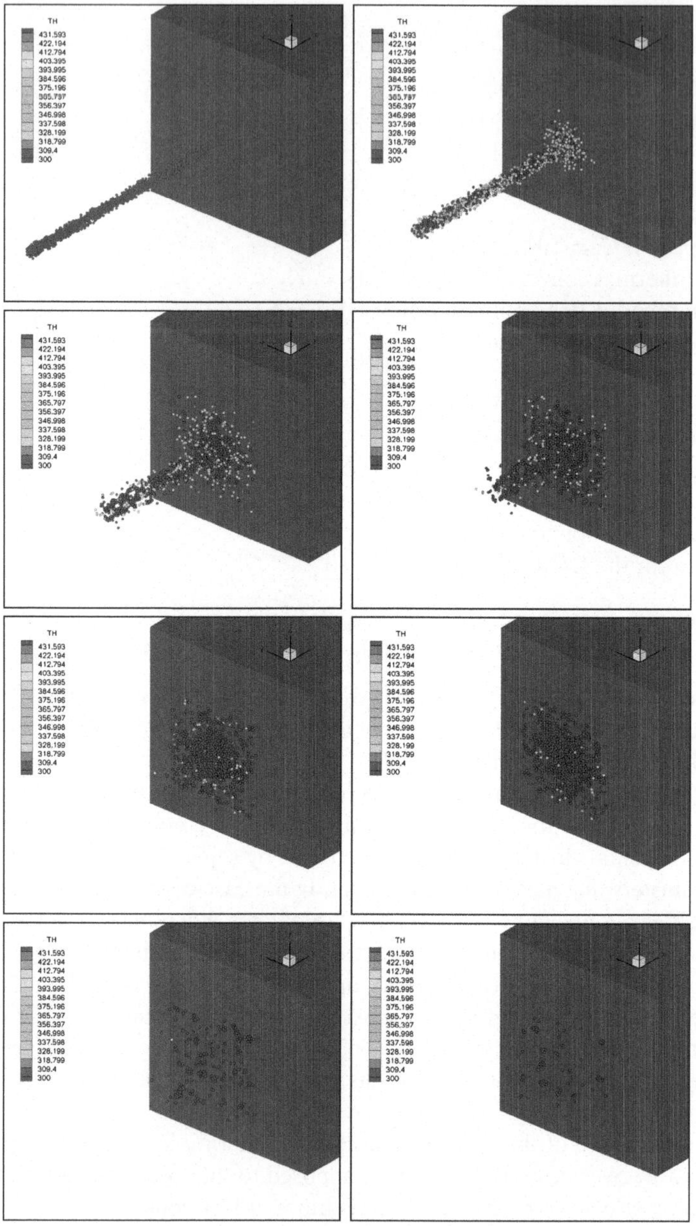

Fig. 3.11 *Top* to *bottom* and *left* to *right*: impact of a charged jet against an immovable obstacle. Note that only particles that are in the domain of interest are shown (Zohdi [110])

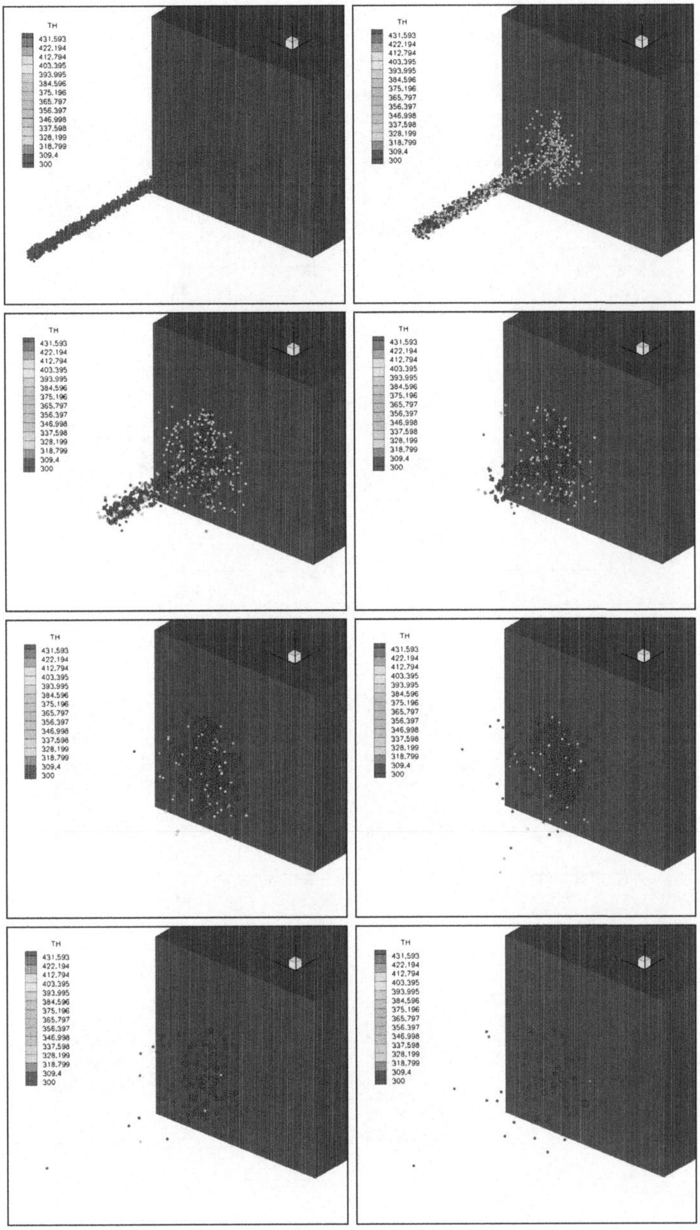

Fig. 3.12 *Top* to *bottom* and *left* to *right*: impact of a charged jet against an electromagnetic field ahead of an immovable obstacle. Note that only particles that are in the domain of interest are shown (Zohdi [110])

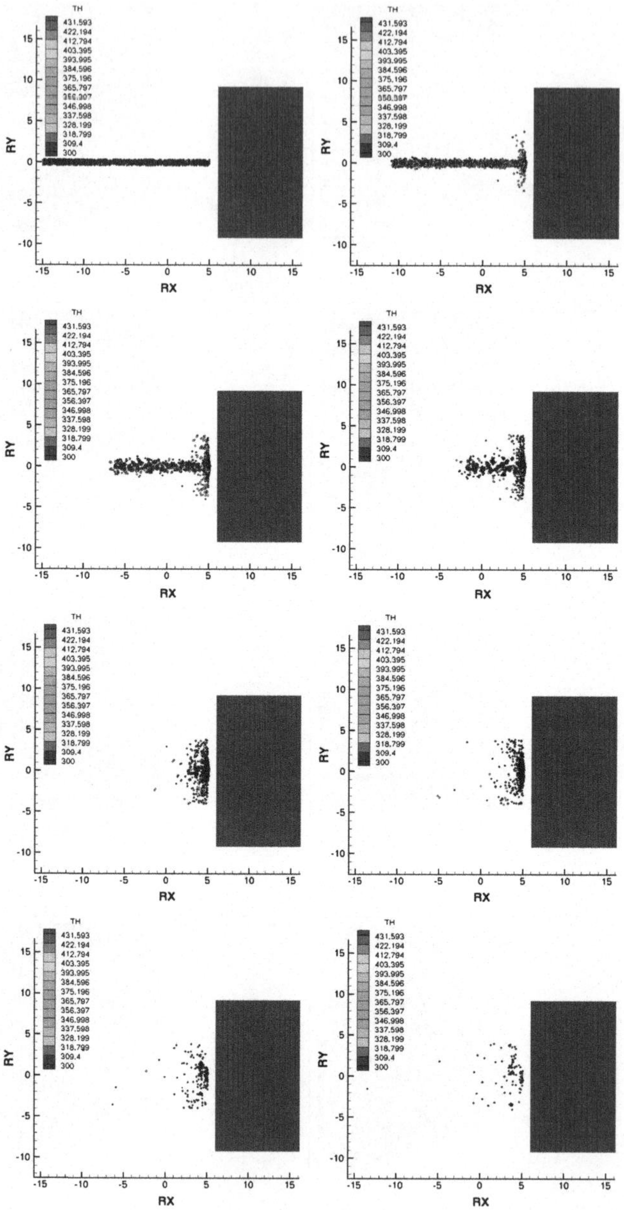

Fig. 3.13 *Top* to *bottom* and *left* to *right* (*side view*): impact of a charged jet against an electromagnetic field ahead of an immovable obstacle. Note that only particles that are in the domain of interest are shown (Zohdi [110])

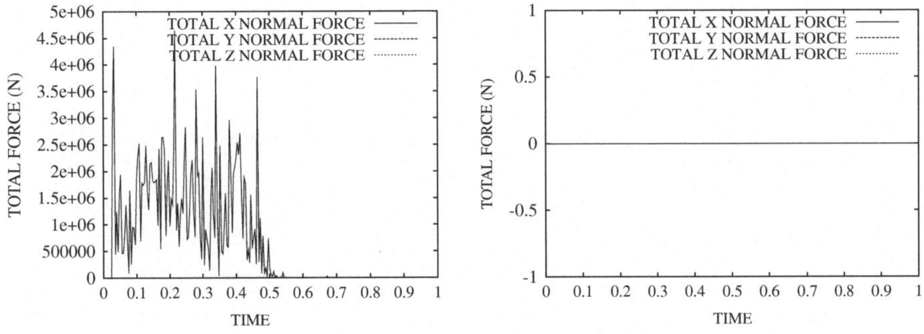

Fig. 3.14 All forces acting on the obstacle. *Left*: charged particles with no electromagnetic fields present and (*right*) charged particles in the presence of an electromagnetic field. No contact is made when the electromagnetic field is present. Note that only particles that are in the domain of interest are included in the computations (Zohdi [110])

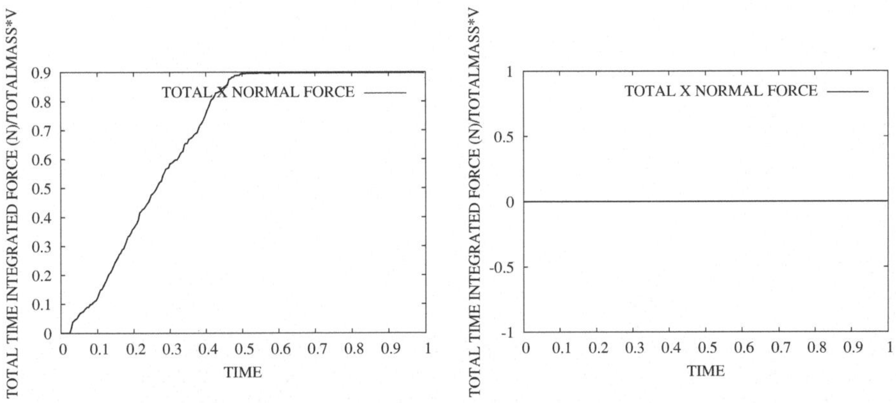

Fig. 3.15 The integrated impulses $\int_0^T \boldsymbol{I}\,dt$ maximum force. *Left*: charged particles with no electromagnetic fields present and (*right*) charged particles in the presence of an electromagnetic field. Note that only particles that are in the domain of interest are included in the computations (Zohdi [110])

where $\mathcal{M} = \sum_{i=1}^{N_p} m_i$ is the total system mass

$$\boldsymbol{r}_{cm} \stackrel{\text{def}}{=} \frac{1}{\mathcal{M}} \sum_{i=1}^{N_p} m_i \boldsymbol{r}_i \tag{3.110}$$

and

$$\boldsymbol{v}_{cm} \stackrel{\text{def}}{=} \frac{1}{\mathcal{M}} \sum_{i=1}^{N_p} m_i \boldsymbol{v}_i. \tag{3.111}$$

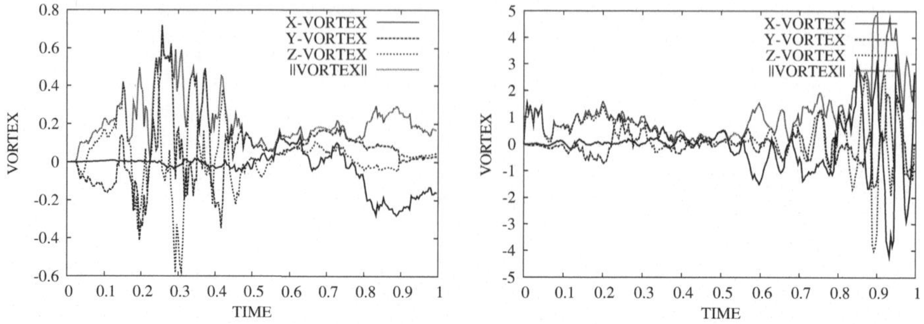

Fig. 3.16 The total vortex. *Left*: charged particles with no electromagnetic fields present and *right*: charged particles in the presence of an electromagnetic field. The Lorentz force induces a large amount of spin within the system, relative to the case where no electromagnetic field present. Note that only particles that are in the domain of interest are included in the computations (Zohdi [110])

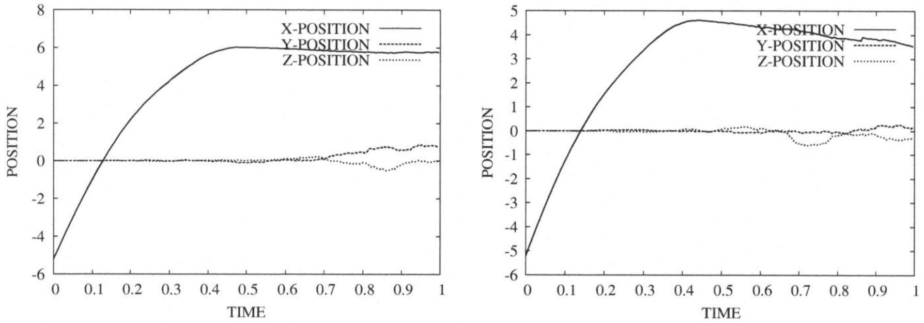

Fig. 3.17 The CG-position. *Left*: charged particles with no electromagnetic fields present and (*right*) charged particles in the presence of an electromagnetic field. Note that only particles that are in the domain of interest are included in the computations (Zohdi [110])

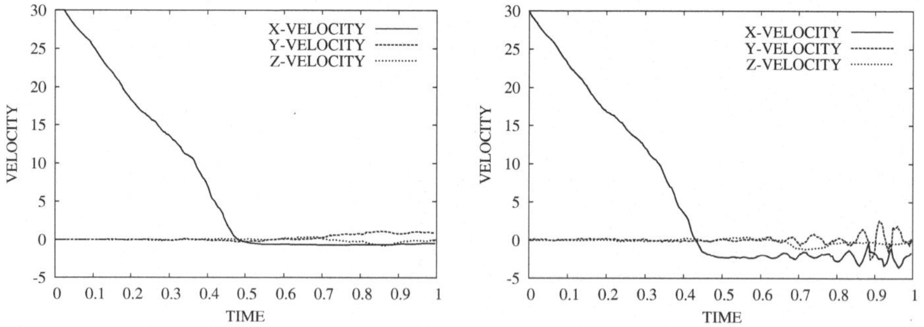

Fig. 3.18 The CG-velocity. *Left*: charged. *Right*: charged with field (note the effects is helical-type spin). Note that only particles that are in the domain of interest are included in the computations (Zohdi [110])

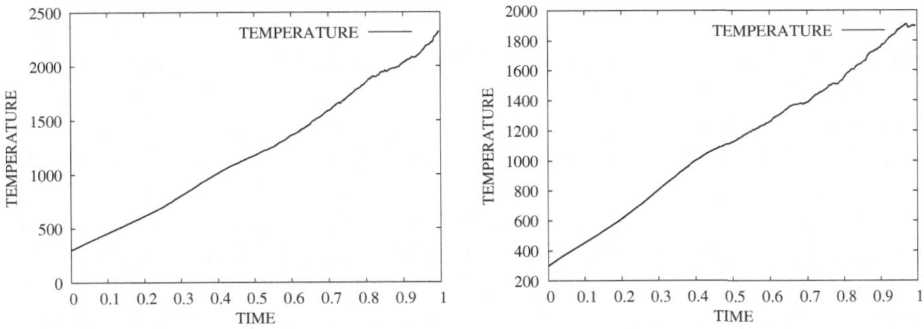

Fig. 3.19 The temperature of the particles within the box. *Left*: charged particles with no electromagnetic fields present and (*right*) charged particles in the presence of an electromagnetic field. The temperature is higher in the case where no electromagnetic field is present because of impact against the rigid foundation. Note that only particles that are in the domain of interest are included in the computations (Zohdi [110])

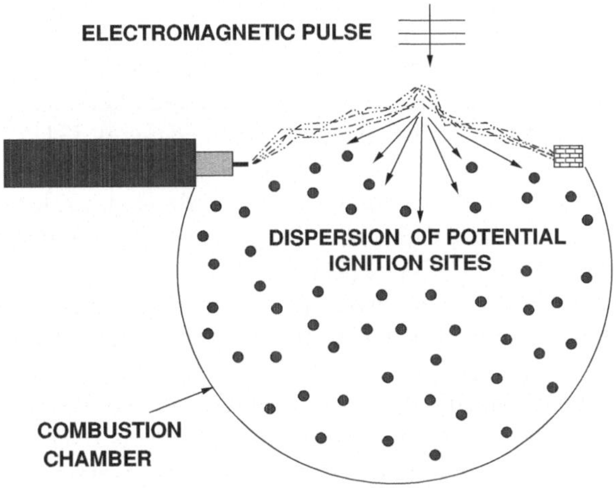

Fig. 3.20 A spark comprised of a charged ion-jet and a possible ablation scheme, with a designer microwave (Zohdi [110])

The discrete vortex is an indication of the spinning of the particles, relative to the system center of mass. In the case of having an electromagnetic wall present, the vortex is significantly higher (Fig. 3.17). *This is due to the Lorentz force* (Fig. 3.10).

3.9 Closing Remarks

We close by commenting on an application not addressed earlier, namely, *modern ignition systems in internal combustion engines*. Currently, the development of ignition systems for ultra-lean fuel mixtures, for example, ethanol, is an area of high interest for spark (a localized electrical ion-jet) control. Because of the advancements in sensor and control systems in harsh environments (see Azevedo et al. [2] and Schwartz [82]), the real-time adjustment of in situ ignition systems has emerged, and is an area of active research. In particular, spark dispersion and control via electromagnetic (microwave) techniques is now possible. Such approaches are important for developing hybrid systems involving Compression Ignition Direct Injection (CIDI) and Homogeneous Charge Compression Ignition (HCCI) engine platforms in order to improve the performance of a broad class of engines. Such systems can lead to improved efficiency by igniting ultra-lean fuel mixtures at low temperatures where standard compression engines are limited and misfiring occurs. In particular, it has recently been proposed that shortly after a spark (charged plasma) is discharged into the combustion chamber, a microwave is released to disperse the spark throughout the combustion chamber. Because of the dispersion of the spark throughout the combustion chamber, the ignition system relies far less on the propagation of the flame to ignite the fuel and, hence, far leaner fuel mixtures may be employed without misfire occurring. Until recently, the detailed control of sparks has been unreachable. However, within the last decade, microtechnology, namely, microelectronic sensors, has made it possible for engineers to control combustion events via the propagation of sparks and flame fronts. This is a large field, and we list a cross-section of that research for the interested reader: Ikeda et al. [31], Kaneko et al. [35], Liepold et al. [45], Phelps [70], Aleiferis et al. [1], Johansson [34], Kogoma [42], Phuoc [76], Morsy et al. [59], Morsy et al. [60] Ma et al. [52], Ma et al. [53], Mohamed et al. [58], Weinberg and Wilson [88], Dale et al. [13], Ronney [79], Beduneau [9], Chen and Lewis [11], Phuoc [77], Kim et al. [40], Ombrello and Ju [64], Mintoussov et al. [57], Korolev and Matveev [43], Kim, et al. [40], Esakov, et al. [21], Linkenheil et al. [50], Linkenheil et al. [51], Ikeda et al. [32], Kawahara et al. [37], Mehresh et al. [55], Bogin et al. [10] and Prager et al. [75]. A further understanding of what influences localized electrical jets is critical for the improvement of combustion processes. In many cases, the analysis of such sprays/jets require the simulation of the electromagnetic response, as well as its resulting coupled thermal response, which can be important to determine possible "hot spots" in the combusting system and to avoid nonuniform dispersion. The current work of the author involves the development of models and numerical solution strategies to analyze the coupled response of such sprays/sparks in order to facilitate the creation of "designer sparks."

An important aspect of any model is identification of parameters which force the system behavior to match a (desired) target response. For example, in the ideal case, one would like to determine the type of system parameters that produce certain overall responses, via numerical simulations, in order to guide or minimize time-consuming laboratory tests. A relatively straightforward way of achieving this is to consider inverse problems whereby particulate parameters are sought which deliver a desired overall behavior by minimizing a cost function. For example, a design problem can be established by defining an N-tuple design vector, denoted by $\boldsymbol{\Lambda} \stackrel{\text{def}}{=} (\Lambda_1, \Lambda_2, \ldots, \Lambda_N)$, for example, consisting of the following free system variables: (1) Particulate masses, (2) Particulate volume fraction, (3) Particulate interaction laws, (4) Particulate velocities, (5) Particulate friction and (6) External electromagnetic fields to minimize particulate agglomeration and maximize particulate dispersion. Generally, formulations of such objective functions will possess nonconvex and nondifferentiable dependency on the design variables (especially if there are constraints).[14] The minimization of such objective functions can be achieved by using a two stage approach whereby (1) one determines promising optimal regions in parameter space using (nonderivative) algorithms (for example, evolutionary "genetic" algorithms, simulated annealing, etc.) and then (2) one applies classical gradient-based schemes in locally convex regions if the objective function is smooth enough (since such approaches (Figs. 3.15, 3.16, 3.18, 3.20) are generally extremely efficient for the minimization of smooth convex functions). There are a variety of nonderivative algorithms that employ concepts of species evolution, for example, reproduction, mutation and crossover. Such methods can be traced back, at least, to the work of Holland [29]. For reviews of such methods, see Goldberg [26], Davis [14], Onwubiko [63], Kennedy and Eberhart [38], Lagaros et al. [44], Papadrakakis et al. [65–68], and Goldberg and Deb [27]. Such approaches have been extensively applied to related problems involving randomly dispersed particulates in continuous media in Zohdi [95–113].

References

1. Aleiferis, P. G., Taylor, A. M. K. P., Whitelaw, J. H., Ishii, K., & Urata, Y. (2000). Cyclic variations of initial flame kernel growth in a honda VTEC-E lean-burn spark- ignition engine. SAE Paper No. 2000–01-1207.
2. Azevedo, R. G., Jones, D. G., Jog, A. V., Jamshidi, B., Myers, D. R., Chen, L., et al. (2007). A SiC MEMS resonant strain sensor for harsh environment applications. *IEEE Sensors Journal, 7*(4), 568–576.
3. Armero, F., & Simo, J. C. (1992). A new unconditionally stable fractional step method for non-linear coupled thermomechanical problems. *International Journal of Numerical Methods in Fluids, 35*, 737–766.
4. Armero, F., & Simo, J. C. (1993). A-priori stability estimates and unconditionally stable product formula algorithms for non-linear coupled thermoplasticity. *International Journal of Plasticity, 9*, 9149–1820.
5. Armero, F., & Simo, J. C. (1996). Formulation of a new class of fractional-step methods for the incompressible MHD equations that retains the long-term dissipativity of the continuum dynamical system, integration algorithms for classical mechanics. *The Fields Institute Communications, 10*, 1–23.
6. Armero, F. (1999). Formulation and finite element implementation of a multiplicative model of coupled poro-plasticity at finite strains under fully saturated conditions. *Computer Methods in Applied Mechanics and Engineering, 171*, 205–241.
7. Bathe, K. J. (1996). *Finite element procedures*. Englewood: Prentice-Hall.
8. Becker, E. B., Carey, G. F., & Oden, J. T. (1980). *Finite elements: An introduction*. Englewood: Prentice-Hall.

[14] Furthermore, such objective functions are usually "noisy," i.e., they possess a stochastic nature, partially due to the random particulate microstructure and sample size.

9. Beduneau, J. L., Kim, B., Zimmer, L., & Ikeda, Y. (2003). Measurements of minimum ignition energy in premixed laminar methane/air flow by using laser induced spark. *Combustion and Flame, 132*, 653–665.
10. Bogin, G., Chen, J. Y., & Dibble, R. W. (2008). The effects of intake pressure, fuel concentration, and bias voltage on the detection of ions in a homogeneous charge compression ignition (HCCI) engine. Proceedings of the Combustion Institute, 32.
11. Chen, Y. L., & Lewis, J. W. L. (2001). Visualisation of laser-induced breakdown and ignition. *Optics Express, 9*(7), 360–372.
12. Cho, H., & Barber, J. R. (1999). Stability of the three-dimensional coloumb friction law. *Proceedings of the Royal Society, 455*(1983), 839–862.
13. Dale, J. D., Smy, P. R., & Clements, R. M. (1978). Laser ignited internal combustion engine — an experimental study. SAE-780329, Detroit.
14. Davis, L. (1991). *Handbook of genetic algorithms*. Boston: Thompson Computer Press.
15. Doltsinis, I. St. (1993). Coupled field problems-solution techniques for sequential and parallel processing. In M. Papadrakakis (Ed.), Solving large-scale problems in mechanics. New York:Wiley.
16. Doltsinis, I. St. (1997). Solution of coupled systems by distinct operators. *Engineering Computations, 14*, 829–868.
17. Donev, A., Cisse, I., Sachs, D., Variano, E. A., Stillinger, F., Connelly, R., Torquato, S., & Chaikin, P. (2004a). *Improving the density of jammed disordered packings using ellipsoids. Science,13* (303), 990–993.
18. Donev, A., Stillinger, F. H., Chaikin, P. M., & Torquato, S. (2004b). Unusually dense crystal ellipsoid packings. *Physical Review Letters, 92*, 255506.
19. Donev, A., Torquato, S., & Stillinger, F. (2005). Neighbor list collision-driven molecular dynamics simulation for nonspherical hard particles-I. algorithmic details. *Journal of Computational Physics, 202*, 737.
20. Duran, J. (1997). *Sands, powders and grains: An introduction to the physics of Granular Matter*. New York: Springer.
21. Esakov, I. I., Grachev, L. P., Khodataev, K. V., Vinogradov, V. V., & Van Wie, D. M. (2006). Propane-air mixture combustion assisted by MW discharge in a speedy airflow. *IEEE Transactions on Plasma Science, 34*(6), 2497.
22. Farhat, C., Lesoinne, M., & Maman, N. (1995). Mixed explicit/implicit time integration of coupled aeroelastic problems: Three-field formulation, geometric conservation and distributed solution. *International Journal for Numerical Methods in Fluids, 21*, 807–835.
23. Farhat, C., & Lesoinne, M. (2000). Two efficient staggered procedures for the serial and parallel solution of three-dimensional nonlinear transient aeroelastic problems. *Computer Methods in Applied Mechanics and Engineering, 182*, 499–516.
24. Farhat, C., van der Zee, G., & Geuzaine, P. (2006). Provably second-order time-accurate loosely-coupled solution algorithms for transient nonlinear computational aeroelasticity. *Computer Methods in Applied Mechanics and Engineering, 195*, 1973–2001.
25. Fish, J., & Chen, W. (2003). Modeling and simulation of piezocomposites. *Computer Methods in Applied Mechanics and Engineering, 192*, 3211–3232.
26. Goldberg, D. E. (1989). Genetic algorithms in search, optimization and machine learning. Reading: Addison-Wesley.
27. Goldberg, D. E., & Deb, K. (2000). Special issue on genetic algorithms. *Computer Methods in Applied Mechanics and Engineering, 186*(2–4), 121–124.
28. Goldsmith, W. (2001). *Impact: The theory and physical behavior of colliding solids*. Toronto: Dover Re-issue.
29. Holland, J. H. (1975). *Adaptation in natural and artificial systems*. Ann Arbor: University of Michigan Press.
30. Hughes, T. J. R. (1989). *The finite element method*. New York: Prentice Hall.
31. Ikeda, Y., Nishiyama, A., & Kaneko, M. (2009). Microwave enhanced ignition process for fuel mixture at elevated pressure of 1 MPa. 47th AIAA Aerospace Sciences Meeting Including the New Horizons Forum and Aerospace Exposition, 5–8 January 2009, Orlando: Kluwer Academic / Plenum Publishers.

32. Ikeda, Y., Nishiyama, A., Kawahara, N., Tomita, E., & Nakayama, T. (2006). Local equivalence ratio measurement of CH4/air and C3H8/air laminar flames by laser-induced breakdown spectroscopy, *44th AIAA Aerospace Sciences Meeting and Exhibit*, 9–12 January 2006, Reno, Nevada, AIAA Paper No.2006-965, 2006.
33. Johnson, K. (1985). *Contact mechanics*. Cambridge: Cambridge University Press.
34. Johansson, B. (1996). Cycle to cycle variations in SI engines — the effects of fluid flow and gas composition in the vicinity of the spark plug on early combustion. SAE Paper 962084.
35. Kaneko, M., Nishiyama, A., Jeong, H., Kantano, H., & Ikeda, I. (2008). Combustion characteristics of microwave plasma combustion engine. Japanese Society of Automotive Engineers, 7–11.
36. Kansaal, A., Torquato, S., & Stillinger, F. (2002). Diversity of order and densities in jammed hard-particle packings. *Physical Review E, 66*, 041109.
37. Kawahara, K., Ueda, K., & Ando, H. (1998). *Mixing control strategy for engine performance improvement in a gasoline direct-injection engine*. SAE Paper No. 980158.
38. Kennedy, J., & Eberhart, R. (2001). *Swarm intelligence*. San Francisco: Morgan Kaufmann Publishers.
39. Kikuchi, N., & Oden, J. T. (1988). *Contact problems in elasticity: A study of variational inequalities and finite element methods*. Philadelphia: SIAM.
40. Kim, Y., Ferreri, V. W., Rosocha, L. A., Anderson, G. K., Abbate, S., & Kim, K.-T. (2006). Effect of plasma chemistry on activated propane/air flames. *IEEE Transactions on Plasma Science, 34*(6), 2532–2536.
41. Klarbring, A. (1990). Examples of nonuniqueness and nonexistence of solutions to quasistatic contact problems with friction. *Ingenieur-Archiv, 60*, 529–541.
42. Kogoma, M. (2003). Generation of atmospheric-pressure glow and its applications. *Journal of Plasma and Fusion Research, 79*(10), 1000.
43. Korolev, Y. D., & Matveev, I. B. (2006). Nonsteady-state processes in a plasma pilot for ignition and flame control. *IEEE Transactions on Plasma Science, 34*(6), 2507.
44. Lagaros, N., Papadrakakis, M., & Kokossalakis, G. (2002). Structural optimization using evolutionary algorithms. *Computers and Structures, 80*, 571–589.
45. Leipold, F., Stark, R. H., El-Habachi, A., & Schoenbach, K. H. (2000). Electron density measurements in an atmospheric pressure air plasma by means of IR heterodyne interferometry. *Journal of Physics D: Applied Physics, 33*, 2268–2273.
46. Lesoinne, M., & Farhat, C. (1998). Free staggered algorithm for nonlinear transient aeroelastic problems. *AIAA Journal, 36*(9), 1754–1756.
47. Le Tallec, P., & Mouro, J. (2001). Fluid structure interaction with large structural displacements. *Computer Methods in Applied Mechanics and Engineering, 190*(24–25), 3039–3067.
48. Lewis, R. W., & Schrefler, B. A. (1998). *T*he finite element method in the static and dynamic deformation and consolidation of porous media (2nd ed.). New York: Wiley Press.
49. Lewis, R. W., Schrefler, B. A., & Simoni, L. (1992). Coupling versus uncoupling in soil consolidation. *International Journal for Numerical and Analytical Methods in Geomechanics, 15*, 533–548.
50. Linkenheil, K., Ruoss, R. O., & Heinrich, W. (2004). Design and evaluation of a novel spark-plug based on a microwave coaxial resonator. *3*4th European Microwave Conference, 3(11–15), 1561–1564.
51. Linkenheil, K., Ruoss, H. O., Grau, T., Seidel, J., & Heinrich, W. (2005). A novel spark-plug for improved ignition in engines with gasoline direct injection (GDI). *IEEE Transactions on Plasma Science,33*(5), 1696.
52. Ma, J. X., Alexander, D. R., & Poulain, D. E. (1998). Laser spark ignition and combustion characteristics of methane-air mixtures. *Combustion and Flame, 112*, 492–506.
53. Ma, J. X., Ryan, T. W., & Buckingham, J. P. (1998). Nd:YAG laser ignition of natural gas", ASME, 98-ICE-114.
54. Martins, J. A. C., & Oden, J. T. (1987). Existence and uniqueness results in dynamics contact problems with nonlinear normal and friction interfaces. *Nonlinear Analysis, 11*, 407–428.

55. Mehresh, P., Souder, J., Flowers, D., Riedel, U., & Dibble, R. W. (2005). Combustion timing in HCCI engines determined by ion-sensor: Experimental and kinetic modeling. *Proceedings of the Combustion Institute, 30*, 2701–2709.
56. Michopoulos, G., Farhat, C., & Fish, J. (2005). Survey on modeling and simulation of multiphysics systems. *Journal of Computing and Information Science in Engineering, 5*(3), 198–213.
57. Mintoussov, E., Anokhin, E., & Starikovskii, A. Y. (2007). Plasma-assisted combustion and fuel reforming. 45th AIAA Aerospace Sciences Meeting and Exhibit, Jan. 8–11, 2007, Reno, Nevada, AIAA Paper no. 2007–1382.
58. Mohamed, A. H., Block, R., & Schoenbach, K. H. (2002). Direct current discharges in atmospheric air. *IEEE Transactions on Plasma Science, 30*(1), 182–183.
59. Morsy, M. H., Ko, Y. S., Chung, S. H., & Cho, P. (2001). Laser-induced two point ignition of premixture with a single-shot laser. *CoFl., 125*, 724–727.
60. Morsy, M. H., & Chung, S. H. (2003). Laser induced multi-point ignition with a single-shot laser using two conical cavities for hydrogen/air mixtures. *Experimental Thermal and Fluid Science, 27*, 491–497.
61. Oden, J. T., & Pires, E. (1983). Nonlocal and nonlinear friction laws and variational principles for contact problems in elasticity. *Journal of Applied Mechanics, 50*, 67–76.
62. Tuzun, U., & Walton, O. R. 1992. Micro-Mechanical modeling of load dependent friction in contacts of elastic spheres. Journal of Physics D: Applied Physics, 25(1A), A44–A52.
63. Onwubiko, C. (2000). *Introduction to engineering design optimization*. Upper Saddle River: Prentice Hall.
64. Ombrello, T., & Ju, Y. (2007). Ignition enhancement using agnetic gliding arc. 45th AIAA Aerospace Sciences Meeting and Exhibit, Jan. 8–11, Reno, Nevada, AIAA Paper no. 2007–1025.
65. Papadrakakis, M., Lagaros, N., Thierauf, G., & Cai, J. (1998). Advanced solution methods in structural optimisation using evolution strategies. *Engineering Computations Journal, 15*(1), 12–34.
66. Papadrakakis, M., Lagaros, N., & Tsompanakis, Y. (1998). Structural optimization using evolution strategies and neutral networks. *Computer Methods in Applied Mechanics, 156*(1), 309–335.
67. Papadrakakis, M., Lagaros, N., & Tsompanakis, Y. (1999). Optimization of large-scale 3D trusses using evolution strategies and neural networks. *International Journal of Space Structures, 14*(3), 211–223.
68. Papadrakakis, M., Tsompanakis, J., & Lagaros, N. (1999). Structural shape optimisation using evolution strategies. *Engineering Optimization, 31*, 515–540.
69. Park, K. C., & Felippa, C. A. (1983). Partitioned analysis of coupled systems. In T. Belytschko & T. J. R. Hughes (Eds.), Computational methods for transient Analysis.
70. Phelps, A. V. (1987). Excitation and ionization coefficients. In L. G. Christophourou & D. W. Bouldin (Eds.), *Gaseous Di- electrics V*. New York: Pergamon.
71. Piperno, S. (1997). Explicit/implicit fluid/structure staggered procedures with a structural predictor and fluid subcycling for 2D inviscid aeroelastic simulations. *International Journal for Numerical Methods in Fluids, 25*, 1207–1226.
72. Piperno, S., Farhat, C., & Larrouturou, B. (1995). Partitioned procedures for the transient solution of coupled aeroelastic problems — part I: Model problem, theory, and two-dimensional application. *Computer Methods in Applied Mechanics and Engineering, 124*(1–2), 79–112.
73. Piperno, S., & Farhat, C. (2001). Partitioned procedures for the transient solution of coupled aeroelastic problems — part II: Energy transfer analysis and three-dimensional applications. *Computer Methods in Applied Mechanics and Engineering, 190*, 3147–3170.
74. Pöschel, T., & Schwager, T. (2004). *Computational granular dynamics*. Berlin: Springer.
75. Prager, J., Riedel, U., & Warnatz, J. (2007). Modeling ion chemistry and charged species diffusion in lean methane-oxygen flames. *Proceedings of the Combustion Institute, 31*(1), 1129–1137.

76. Phuoc, T. (2000). Single-point versus multi-point laser ignition: Experimental measurements of combustion times and pressures. *CoFl., 122*, 508–510.
77. Phuoc, T. (2000). Laser spark ignition: Experimental determination of laser-induced breakdown thresholds of combustion gases. *Optics Communication, 175*, 419–423.
78. Rietema, K. (1991). *Dynamics of fine powders*. New York: Springer.
79. Ronney, P. D. (1994). Laser versus conventional ignition of flames. *Optical Engineering, 33*(2), 510.
80. Schmidt, L. (1998). *The engineering of chemical reactions*. New York: Oxford University Press.
81. Schrefler, B. A. (1985). A partitioned solution procedure for geothermal reservoir analysis. *Communications in Applied Numerical Methods, 1*, 53–56.
82. Schwartz, S. W., Myers, D. R., Kramer, R. K., Choi, S., Jordan, A., Wijesundara, M. B. J., Hopcroft, M. A., & Pisano, A. P. (2008). Silicon and silicon carbide survivability in an in-cylinder combustion environment. PowerMEMS 2008, Sendai, Japan, Nov 9–12.
83. Sukop, M. C., & Thorne, D. T. (2006). *Lattice-Boltzmann Modeling: An introduction for geoscientists and engineers*. Berlin: Springer.
84. Szabo, B., & Babúska, I. (1991). *Finite element analysis*. New York: Wiley Interscience.
85. Tabor, D. (1975). Interaction between surfaces: Adhesion and friction. In Blakely, J. M. (ed.), Surface physics of materials, (vol. II, Chap. 10). New York: Academic Press.
86. Torquato, S. (2002). *Random heterogeneous materials: Microstructure and macroscopic properties*. New York: Springer.
87. Turska, E., & Schrefler, B. A. (1994). On consistency, stability and convergence of staggered solution procedures. *Rendiconti Matematico Account, 5*(9), 265–271.
88. Weinberg, F. J., & Wilson, J. R. (1971). A preliminary investigation of the use of focused laser beams for minimum ignition energy studies. *Proceedings of the Royal Society London, A321*, 41–52.
89. Widom, B. (1966). Random sequential addition of hard spheres to a volume. *Journal of Chemical Physics, 44*, 3888–3894.
90. Wriggers, P. (2002). *Computational contact mechanics*. Chichester: Wiley.
91. Wriggers, P. (2008). *Nonlinear finite element analysis*. Berlin: Springer.
92. Zienkiewicz, O. C. (1984). Coupled problems and their numerical solution. In R. W. Lewis, P. Bettes, & E. Hinton (Eds.), *Numerical methods in coupled systems* (pp. 35–58). Chichester: Wiley.
93. Zienkiewicz, O. C., Paul, D. K., & Chan, A. H. C. (1988). Unconditionally stable staggered solution procedure for soil-pore fluid interaction problems. *International Journal for Numerical Methods in Engineering, 26*, 1039–1055.
94. Zienkiewicz, O. C., & Taylor R. L. (1991). *The finite element method* (vols. I and II). New York: McGraw-Hill.
95. Zohdi, T. I. (2002). An adaptive-recursive staggering strategy for simulating multifield coupled processes in microheterogeneous solids. *International Journal of Numerical Methods in Engineering, 53*, 1511–1532.
96. Zohdi, T. I. (2003). Genetic design of solids possessing a random-particulate microstructure. *PTRS: Mathematical, Physical, and Engineering Sciences, 361*(1806), 1021–1043.
97. Zohdi, T. I. (2003). On the compaction of cohesive hyperelastic granules at finite strains. *Proceedings of the Royal Society, 454*(2034), 1395–1401.
98. Zohdi, T. I. (2003). Computational design of swarms. *International Journal for Numerical Method in Engineering, 57*, 2205–2219.
99. Zohdi, T. I. (2003). Constrained inverse formulations in random material design. Computer Methods in Applied Mechanics and Engineering, 1–20, 192(28–30),18, 3179–3194.
100. Zohdi, T. I. (2004). Modeling and simulation of a class of coupled thermo-chemo-mechanical processes in multiphase solids. *Computer Methods in Applied Mechanics and Engineering, 193*(6–8), 679–699.
101. Zohdi, T. I. (2004). Modeling and direct simulation of near-field granular flows. *The International Journal of Solids and Structures, 42*(2), 539–564.

102. Zohdi, T. I. (2004). A computational framework for agglomeration in thermo-chemically reacting granular flows. *Proceedings of the Royal Society, 460*(2052), 3421–3445.
103. Zohdi, T. I. (2005). Charge-induced clustering in multifield particulate flow. *International Journal for Numerical Method in Engineering, 62*(7), 870–898.
104. Zohdi, T. I. (2006). Computation of the coupled thermo-optical scattering properties of random particulate systems. *Computer Methods in Applied Mechanics and Engineering, 195*, 5813–5830.
105. Zohdi, T. I. (2007). Computation of strongly coupled multifield interaction in particle-fluid systems. *Computer Methods in Applied Mechanics and Engineering, 196*, 3927–3950.
106. Zohdi, T. I. (2007). Particle collision and adhesion under the influence of near-fields. *Journal of Mechanics of Material and Structures,* 2(6), 1011–1018.
107. Zohdi, T. I. (2008). On the computation of the coupled thermo-electromagnetic response of continua with particulate microstructure. *International Journal for Numerical Method in Engineering, 76*, 1250–1279.
108. Zohdi, T. I. (2009). Mechanistic modeling of swarms. *Computer Methods in Applied Mechanics and Engineering, 198*(21–26), 2039–2051.
109. Zohdi, T. I. (2010). Charged wall-growth in channel-flow. *International Journal of Engineering Science, 48*, 15–20.
110. Zohdi, T. I. (2010). On the dynamics of charged electromagnetic particulate jets. *Archives of Computational Methods in Engineering, 17*(2), 109–135.
111. Zohdi, T. I. (2011). Dynamics of clusters of charged particulates in electromagnetic fields. *International Journal for Numerical Method in Engineering, 85*, 1140–1159.
112. Zohdi, T. I. (2007). Introduction to the modeling and simulation of particulate flows. Berkeley: SIAM (Society for Industrial and Applied Mathematics).
113. Zohdi, T. I., & Wriggers, P. (2008). Introduction to computational micromechanics. (second reprinting). Berlin: Springer.

Chapter 4
Charged Particle Impact on Electrified Surfaces

Here, we follow a recent analysis of Zohdi [4], the impact of particles on electrified surfaces has wide-ranging applications, for example in coating technologies. A single particle impacting a surface will rebound according to the basic laws governing a conservation of linear momentum, in conjunction with restitution relations and friction laws. However, if one projects a collection of particles, for example in the form of a droplet, at a surface, collisions between rebounding particles and incoming particles confine the mobility of rebounding particles, forcing them to move tangentially to the surface (Fig. 4.1). Our concern in this chapter is the behavior of collections of charged particles in the form of aggregate particulate-droplets, and the resulting impact behavior on electrified surfaces. The chapter develops expressions for the analytical qualitative trends and large-scale simulation techniques for quantitative information. Specifically, the coherency of the resulting impacting droplet is of interest, which has wide-ranging applications in areas such as inkjet printing, sprays, coatings, etc. A central issue is the determination of key system parameters that lead to a coherent surface, versus a drop that will break apart. The present analysis is partially motivated by coating technologies (for example epitaxy and electrostatic painting) that ionize particulates and electrify surfaces to capture the particles, in order to enhance the quality of coating. Relative to a charge-free and electric-field-free system, the resulting coatings can be properly controlled, provided the system parameters are appropriately set. In this chapter, we address some of the issues involved with modeling and simulation of this type of system.

Remark There are a variety of charged-coating methods, for example:

- Post-atomization charging—whereby the particles come into contact with an electrostatic field downstream of the outlet nozzle. The electrostatic field is produced by electrostatic induction or by electrodes.
- Direct charging—whereby an electrode is immersed in the coating supply.
- Tribological charging—whereby the friction in the nozzle induces an electrostatic charge on the particles as they rub the surface.

T. I. Zohdi, *Dynamics of Charged Particulate Systems*, SpringerBriefs in Applied Sciences and Technology, DOI: 10.1007/978-3-642-28519-6_4,

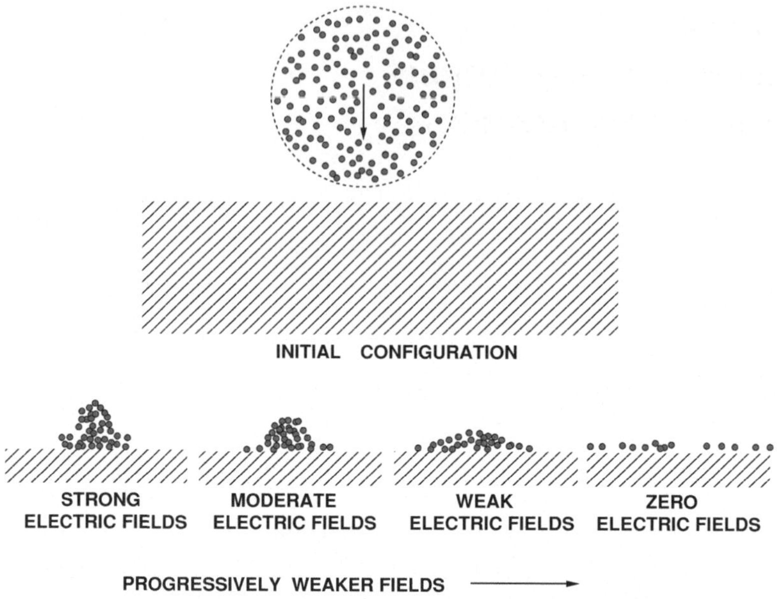

Fig. 4.1 The variation of drop behavior as a function of the local electric field (Zohdi [4])

There are a variety of industrial deposition techniques, and we refer the reader to the surveys of the state of the art found in Martin [2] and [3].

4.1 Transverse Motion of Impacting Particles

In order to understand the effects of collisions between incoming particles and rebounding particles, consider the two uncharged particles in Fig. 4.2. A momentum balance in the normal direction for particle i is

$$m_i v_{in}(t) + \int_t^{t+\delta t} I_n \, dt = m_i v_{in}(t + \delta t) \tag{4.1}$$

and for particle j

$$m_j v_{jn}(t) - \int_t^{t+\delta t} I_n \, dt = m_j v_{jn}(t + \delta t), \tag{4.2}$$

and the coefficient of restitution between particles i and j

$$e_{ij} = \frac{v_{in}(t + \delta t) - v_{jn}(t + \delta t)}{v_{jn}(t) - v_{in}(t)}. \tag{4.3}$$

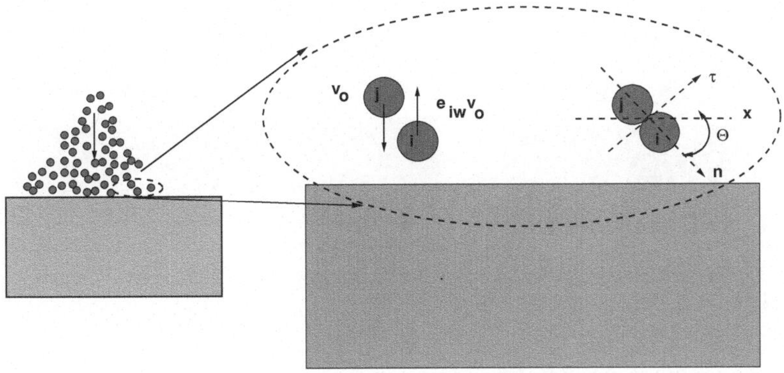

Fig. 4.2 Rebounding collisions inducing lateral motion (Zohdi [4])

For the time being, assuming the contact is frictionless, the tangential velocities are unchanged. Solving the above equations simultaneously yields:

$$v_{in}(t+\delta t) = \left(1+\frac{m_i}{m_j}\right)^{-1}\left(\left(\frac{m_i}{m_j}-e_{ij}\right)v_{in}(t)+\left(1+e_{ij}\right)v_{jn}(t)\right). \tag{4.4}$$

For the rebounding particle, $\boldsymbol{v}_i(t) = e_{iw}v_o\boldsymbol{e}_v$ (e_{iw} is the coefficient of restitution between the particle and wall), thus

$$v_{in}(t) = \boldsymbol{v}_i(t)\cdot\boldsymbol{n} = -e_{iw}v_o sin\Theta, \tag{4.5}$$

while for the incoming particle, $\boldsymbol{v}_j(t) = -v_o\boldsymbol{e}_y$, thus

$$v_{jn}(t) = \boldsymbol{v}_j(t)\cdot\boldsymbol{n} = v_o sin\Theta, \tag{4.6}$$

thus

$$v_{in}(t+\delta t) = \left(1+\frac{m_i}{m_j}\right)^{-1}\left(-\left(\frac{m_i}{m_j}-e_{ij}\right)e_{iw}v_o sin\Theta+(1+e_{ij})v_o sin\Theta\right) \tag{4.7}$$

The post-impact velocity of particle i in the $\boldsymbol{e}_x$ direction is

$$v_{ix_1}(t+\delta t) = v_{in}(t+\delta t)\boldsymbol{n}\cdot\boldsymbol{e}_x+v_{it}(t+\delta t)\boldsymbol{\tau}\cdot\boldsymbol{e}_x = v_{in}(t+\delta t)cos\Theta+v_{it}(t+\delta t)sin\Theta, \tag{4.8}$$

where for simplicity of illustration, we assume frictionless contact, hence, $v_{it}(t+\delta t) = v_{it}(t) = v_o cos\Theta$. Maximizing the horizontal component of the velocity can be determined by

$$\frac{\partial v_{ix_1}(t+\delta t)}{\partial\Theta} = 0 \Rightarrow \Theta^* = \frac{\pi}{4} \tag{4.9}$$

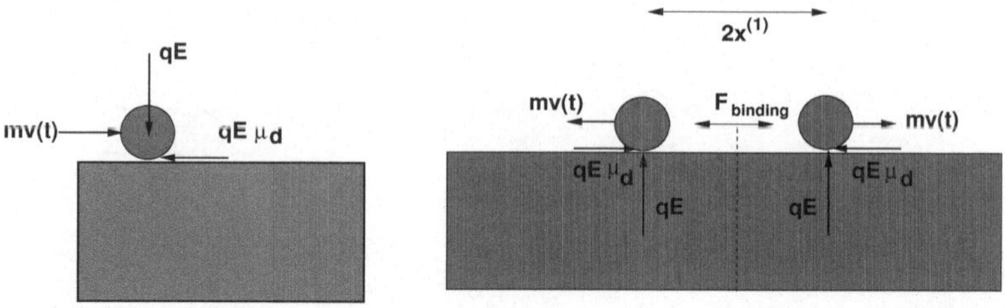

Fig. 4.3 Sliding with electrically-induced friction. *Left* a single particle. *Right* two "bound" particles (Zohdi [4])

and the maximum is, assuming $e_{iw} = e_{ij} = 1$,

$$v_{ix_1}(t + \delta t) = \frac{v_o}{2}. \tag{4.10}$$

The particle j

$$v_{jx_1}(t + \delta t) = -\frac{v_o}{2}. \tag{4.11}$$

The main point is that significant horizontal velocities are generated, pushing particles outwards, as well as propelling particles horizontally back into the flow, producing a cascading effect.

4.2 Qualitative Trends for Charged Particles on Electrified Surfaces

As mentioned, we are concerned with charged droplets formed by particulates when they impact a surface possessing an electric field, in particular the coherency of the resulting "impacted droplet/jet" (Fig. 4.3).

4.2.1 A Single Particle

In order to construct a qualitative estimate of how far charged particles will spread across the surface, let us first consider an isolated particle sliding across the surface. One can determine the travel distance by equating the initial kinetic energy to the energy dissipated by sliding friction, where the normal force is supplied by the electric field acting on the particle's charge qE (Fig. 4.2). This balance yields

$$\underbrace{\frac{1}{2}mv^2(t_o)}_{initial\ kinetic\ energy} = \underbrace{qE\mu_d x(T)}_{energy\ dissipated} \Rightarrow x(T) = \frac{mv^2(t_o)}{2qE\mu_d}. \tag{4.12}$$

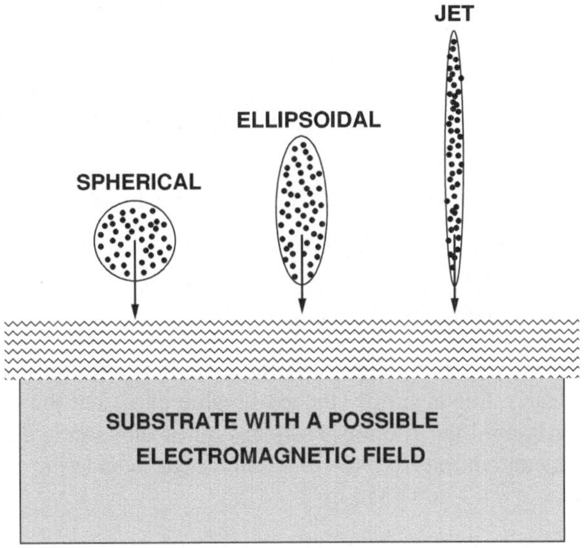

Fig. 4.4 Various classes of droplets

In order to determine the time-scale for this to occur, a momentum balance in the tangential direction yields

$$mv(t_o) - \int_{t_o}^{t_o+T} qE\mu_d \, dt = mv(t_o + T), \tag{4.13}$$

and set $v(t_o + T) = 0$, yielding $T = \frac{mv(t_o)}{qE\mu_d}$.

4.2.2 Two Particles

Now let us repeat the analysis for two bound (charged) particles, with position x_i and x_j, moving in opposite directions (about the symmetry line in Fig. 4.4). For simplicity's sake, let us consider a linear attractive binding of $K(x_i - x_j)$, with binding stiffness K, with corresponding energy $\frac{1}{2}K(x_i - x_j)^2$. Equating the stored, kinetic and dissipated energies, the point where the expansion stops (relative to the symmetry line, where $v = 0$ a maximum) is denoted $x_i = x^{(1)} = -x_j$ is (assuming a spring-like binding law with an unstretched separation (starting) state of $x_i - x_j = 2x^{(0)}$)

$$\begin{aligned}\frac{1}{2}K(2x^{(1)} - 2x^{(0)})^2 &= 2\left(\frac{1}{2}mv^2(t_o) - qE\mu_d(x^{(1)} - x^{(0)})\right)\\ &= mv^2(t_o) - 2qE\mu_d(x^{(1)} - x^{(0)}).\end{aligned} \tag{4.14}$$

Solving for $x^{(1)}$ yields

$$x^{(1)} = x^{(0)} - \frac{qE\mu_d}{2K} \pm \left(\left(\frac{qE\mu_d}{2K} \right)^2 + \frac{mv_o^2}{2K} \right)^{1/2}, \tag{4.15}$$

where the positive root is the physical solution. We note that

- as $qE\mu_d$ increases, $x^{(1)}$ decreases,
- as $mv^2(t_o)$ increases, $x^{(1)}$ increases,
- as K increases, $x^{(1)}$ decreases.

Essentially, this qualitatively indicates how large the droplet will become (expansion). Clearly, this is simply the peak value of x, and the droplet may oscillate back and forth, until the friction dissipates all of the kinetic and stored energy. For example, the contraction from the maximum expansion to the next rest state (where $v = 0$ and $x = x^{(2)}$) is governed by

$$\frac{1}{2}K(2x^{(1)} - 2x^{(0)})^2 = \frac{1}{2}K(2x^{(2)} - 2x^{(0)})^2 + 2qE\mu_d(x^{(1)} - x^{(2)}). \tag{4.16}$$

4.3 Multiple Particulate Examples: Progressively Electrified Surfaces and Figures of Merit for Coating Quality

We utilize the algorithms developed in the previous chapters, and adopt them without alteration, however, we ignore thermal effects in the upcoming analysis. Referring to Fig. 4.5, we consider the following logical figures of merit to assess the deposition process: (using an $x_1 = x, x_2 = y, x_3 = z$ triad (for notational simplicity in this section) with target center at $y = 0, z = 0$)

- Percent deposition in a target (radial) distance: summing the particles that satisfy $\sqrt{y_i^2 + z_i^2} \leq d^{tar}$, and dividing that number by the total number of particles N_p,
- Layer thickness, provided by $\bar{x} = \frac{1}{N_p} \sum x_i$,
- Standard deviation from the mean layer thickness: $S(x_i - \bar{x}) = \frac{1}{N_p} \sum (x_i - \bar{x})^2$,
- Mean distance from the target center: $\bar{d} = \frac{1}{N_p} \sqrt{y_i^2 + z_i^2} \leq R^{tar}$
- Standard deviation from the mean distance: $S(d_i - \bar{d}) = \frac{1}{N_p} \sum (d_i - \bar{d})^2$.

As an example, we considered a group of $N_p = 1000$ randomly positioned particles in a droplet domain (Fig. 4.6). The initial particle radius (monodisperse) was $r_p = 0.05$ m. The absolute dimensions are unimportant for the model problem, and have been normalized. The initial radius of the droplet was set to $R = 1$ initially. All system parameters can be scaled to describe any specific system of interest. The relevant simulation parameters were:

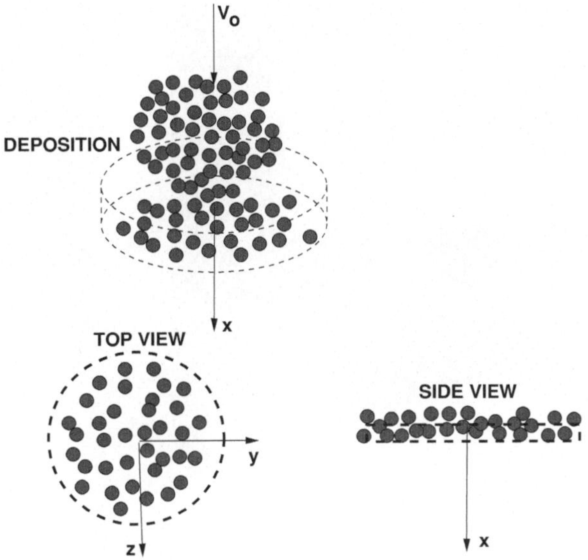

Fig. 4.5 Motivation for the figures of merit (Zohdi [4])

- number of particles $= 1000$, with $\alpha_{ij} = \bar{\alpha}_{ij} q_i q_j c_i c_j$ and $c_i = 1$ (positive), $q_i = q_j = 1$,
- $\bar{\alpha}_{ij1} = 0.5$, $\bar{\alpha}_{ij2} = 0.25$, $\beta_{ij1} = 1$, $\beta_{ij2} = 2$,
- mass density of the particles $= 2000\,\mathrm{kg/m^3}$,
- initial velocity $= (50, 0, 0)$ m/s,
- initial mean position $= (4, 0, 0)$ m,
- coefficient of dynamic friction, $\mu_d = 0.1$,
- coefficient of static friction, $\mu_s = 0.2$,
- baseline coefficient of restitution, $e_o = 0.5$,
- limit of coefficient of restitution, $e^- = 0.2$,
- velocity parameter, $v^* = 10$ m/s,
- target number of fixed point iterations, $K_d = 6$,
- the time-stepping parameter, $\phi = 0.5$,
- $\boldsymbol{E}^{ext} = (E_x^{ext}, 0, 0)$ N/C, which starts after $x = 5$, with wall location at $x = 6$,
- simulation duration $= 1$ s,
- initial time step size $= 0.001$ s,
- time step upper bound $= 0.0025$ s,
- tolerance for the fixed-point iteration $= 10^{-3}$.

Particles that strayed outside of a computational 10×10 m window were thrown out of the computations. Table 4.1 and Fig. 4.6 illustrate the trends with increasing elecric fields, with the following observations:

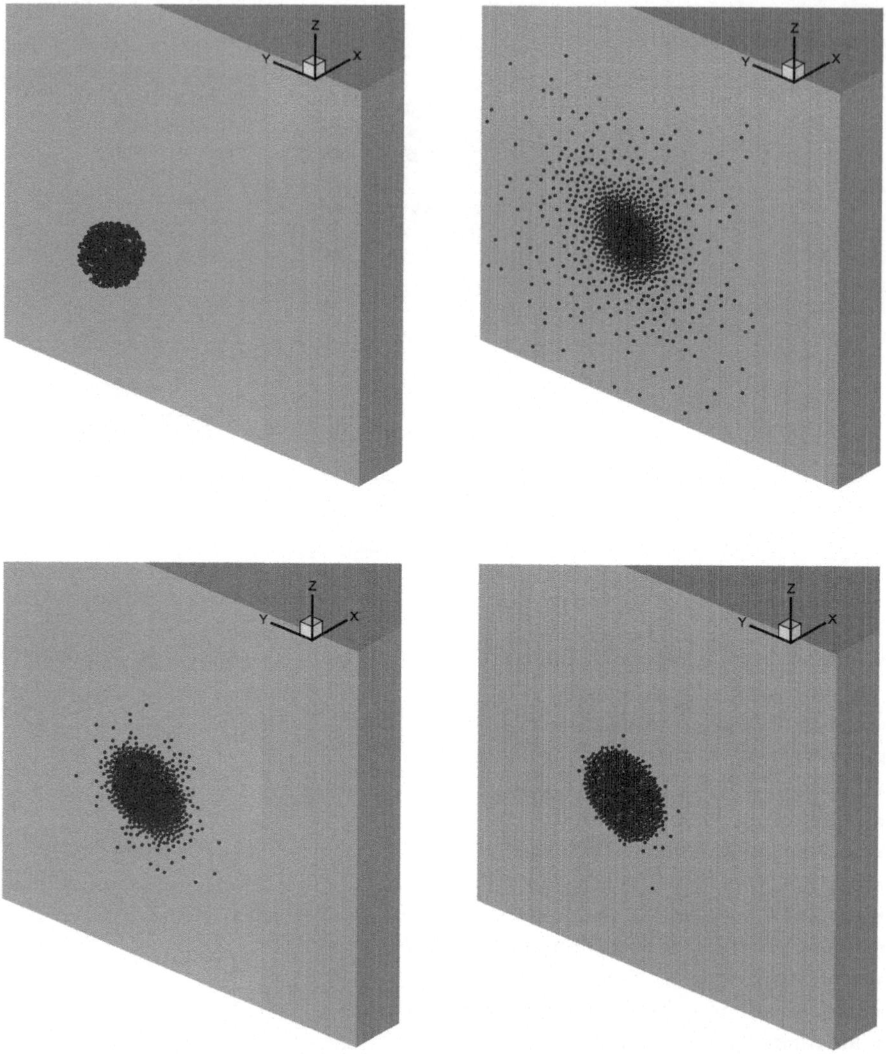

Fig. 4.6 *Top left* the starting configuration. Initially, the cluster-droplet's center is at $x = 4$, the electric field starts at $x = 5$ and the wall is at $x = 6$. The particles each have a radius of $r_p = 0.05$. *Top right* the final configuration of a charged cluster-droplet for $E_x^{ext} = 100\,\text{N/C}$. *Bottom left* for $E_x^{ext} = 1000\,\text{N/C}$. *Bottom right* for $E_x^{ext} = 10000\,\text{N/C}$ (Zohdi [4])

- For the $E_x^{ext} - field \leq 1$ the particles rebound and do not adhere to the surface.
- For the $E_x^{ext} - field \approx 10$ the particles start to adhere with some limited rebounding (a transition scenario),
- For the $E_x^{ext} - field \approx 100$ the particles adhere but spread over the surface incoherently,

Table 4.1 The results as a function of increasing E-fields

$E_x^{ext} - field$	$N_{p,in}/N_{p,tot}$	$\bar{d}$	$S(d_i - \bar{d})$	$\bar{x}$	$S(x_i - \bar{x})$
0.0	0.0630	5.4621	1.8121	4.1127	2.6920
1.0	0.0500	5.6404	1.7289	4.1851	2.7960
10.0	0.1120	5.1276	2.0841	5.5756	1.3868
100.0	0.6040	2.1860	1.7046	5.9498	0.0030
1000.0	0.9480	1.0680	0.5232	5.9500	0.0000
10000.0	0.9950	0.9898	0.4104	5.9504	0.0000

- For the $E_x^{ext} - field \geq 1000$ the particles adhere and form a coherent "splat/dot", with their lateral motion dissipated by friction induced by the electric field contact force.

4.3.1 Remarks on Magnetic Field Interaction

Magnetic fields, due to their production of Lorentz forces, $q_i \boldsymbol{v}_i \times \boldsymbol{B}^{ext}$, acting on each particle i in

$$m_i \dot{\boldsymbol{v}}_i = q_i(\boldsymbol{E}^{ext} + \boldsymbol{v}_i \times \boldsymbol{B}^{ext}), \tag{4.17}$$

will "bend" the overall trajectory of the entire cluster-droplet as show in Fig. 4.7. Qualitative information can be extracted by initially studying a single isolated particle. The governing Eq. 4.17, for component x is

$$\dot{v}_x = \frac{q}{m}(E_x^{ext} + (v_y B_z^{ext} - v_z B_y^{ext})), \tag{4.18}$$

for component y,

$$\dot{v}_y = \frac{q}{m}(E_y^{ext} - (v_x B_z^{ext} - v_z B_x^{ext})), \tag{4.19}$$

and for component z,

$$\dot{v}_z = \frac{q}{m}(E_z^{ext} + (v_x B_y^{ext} - v_y B_x^{ext})). \tag{4.20}$$

In the special case when $\boldsymbol{r}(t = 0) = (r_{x0}, 0, 0)$, $\boldsymbol{v}(t = 0) = (v_{x0}, 0, 0)$, $\boldsymbol{B}^{ext} = B_z^{ext} \boldsymbol{e}_z$ and $\boldsymbol{E}^{ext} = E_x^{ext} \boldsymbol{e}_x$, the solution for the dynamics of an isolated particle is

$$\left\{ \begin{array}{c} v_x(t) \\ v_y(t) \\ v_z(t) \end{array} \right\} = \left\{ \begin{array}{c} v_{x0} cos\omega t - \frac{q\omega E_x^{ext}}{m} sin\omega t \\ -v_{x0} sin\omega t - \frac{q\omega E_x^{ext}}{m} cos\omega t + \frac{q\omega E_x^{ext}}{m} \\ 0 \end{array} \right\} \tag{4.21}$$

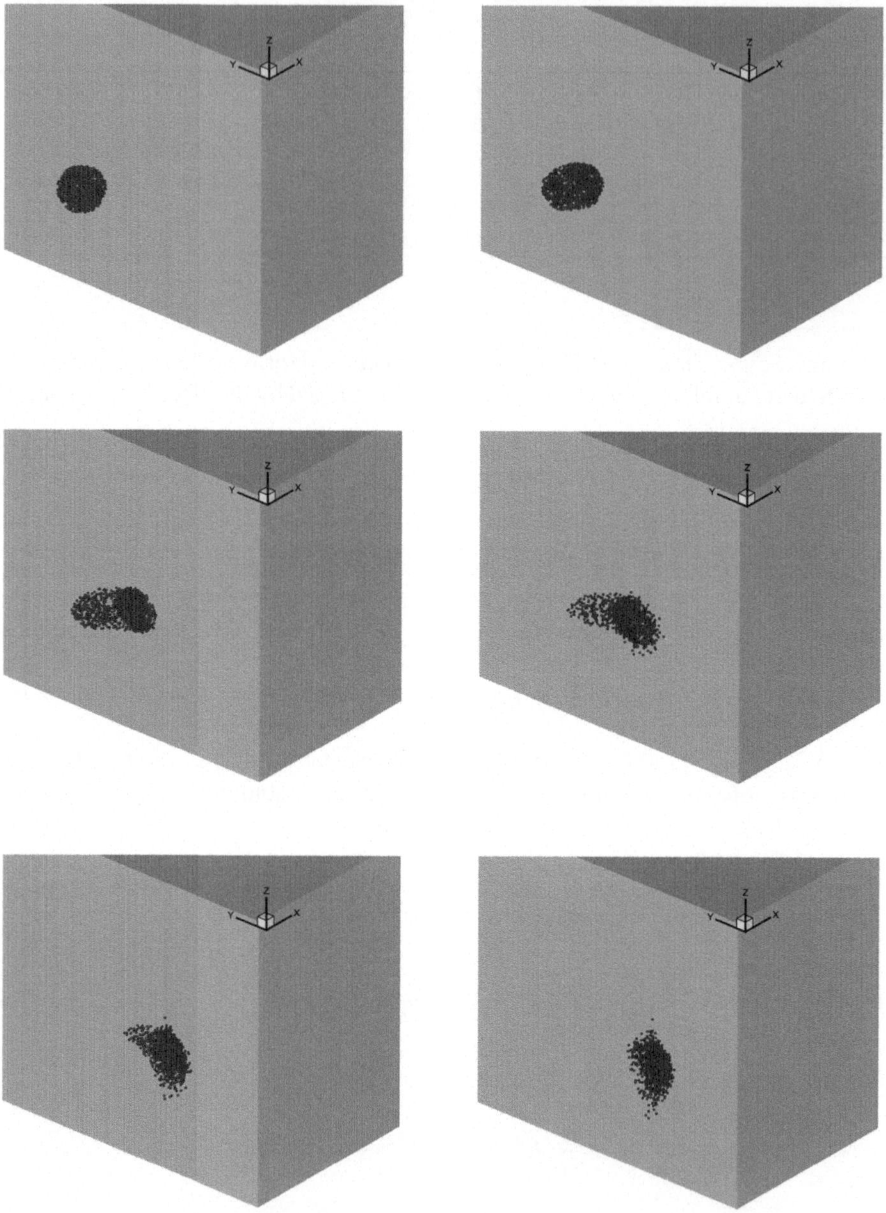

Fig. 4.7 *Left* to *right* and *top* to *bottom* the bending of a drop due to the presence of a magnetic field $\boldsymbol{B}^{ext} = (0, 100, 0)$ N s/C m (also with $\boldsymbol{E}^{ext} = (10000, 0, 0))$ N/C (Zohdi [4])

and

$$\left\{\begin{array}{c} r_x(t) \\ r_y(t) \\ r_z(t) \end{array}\right\} = \left\{\begin{array}{c} \frac{1}{\omega}\left(v_{x0}sin\omega t + \frac{q\omega E_x^{ext}}{m}cos\omega t\right) - \frac{qE_x^{ext}}{m} + r_{x0} \\ \frac{1}{\omega}\left(v_{x0}cos\omega t - \frac{q\omega E_x^{ext}}{m}sin\omega t\right) + \frac{q\omega E_x^{ext}t}{m} - \frac{v_{x0}}{\omega} \\ 0 \end{array}\right\}. \tag{4.22}$$

where $\omega = \frac{qB_z^{ext}}{m}$ is known as the cyclotron frequency. The cyclotron frequency (gyrofrequency) is the angular frequency at which a charged particle makes circular orbits in a plane perpendicular to the static magnetic field. Notice that when $E_x^{ext} = 0$, this traces out the equation of a circle (in the $x-y$ plane) centered at $(r_{x0}, -\frac{v_{x0}}{\omega}, 0)$. The radius of the "magnetically-induced circle" (radius of oscillation) is[1]

$$\mathcal{R} \stackrel{\text{def}}{=} \frac{v_{xO}}{\omega} = \frac{v_{xO}m}{qB_z^{ext}}. \tag{4.23}$$

Thus, if a desired "turning radius" is denoted by $\mathcal{R}$, one may solve for the magnetic field that delivers the desired effect, $B_z^{ext} = \frac{v_{xO}m}{q\mathcal{R}}$. The corresponding time period for one cycle to be completed is $T \stackrel{\text{def}}{=} 2\pi/\omega$.

4.3.2 Observation 1: Higher-Order Statistics

The approach developed provides a fast computational tool to analyze particulate droplets. It can be used on virtually any type of droplet domain. Since the results are derived from a direct numerical simulation, one can also post-process other more detailed statistical information. For example for any quantity of interest, Q (for example the positions of the particles, their velocities, etc.), with a distribution of values (Q_i, i $= 1, 2, \ldots \mathrm{N_p}$ $=$ particles) about an arbitrary reference point, denoted $Q^\star$, as follows:

$$\mathcal{M}_r^{Q_i - Q^\star} \stackrel{\text{def}}{=} \frac{\sum_{i=1}^{N_p} a_i (Q_i - Q^\star)^r}{\sum_{i=1}^{N_p} a_i} \stackrel{\text{def}}{=} \overline{(Q_i - Q^\star)^r}. \tag{4.24}$$

The various moments characterize the distribution, for example:

1. $\mathcal{M}_1^{Q_i - A}$ measures the first deviation from the average, which equals zero,
2. $\mathcal{M}_1^{Q_i - 0} \stackrel{\text{def}}{=} \frac{\sum_{i=1}^{N_p} a_i (Q_i - 0)}{\sum_{i=1}^{N_p} a_i} \stackrel{\text{def}}{=} \overline{(Q_i - 0)} = A$,
3. $\mathcal{M}_2^{Q_i - A}$ is the standard deviation,
4. $\mathcal{M}_3^{Q_i - A}$ is the skewness, which measures the bias, or asymmetry of the distribution of data and

[1] This field generates helical-like motion in three dimensions when $\boldsymbol{E}_x^{ext} \neq \mathbf{0}$.

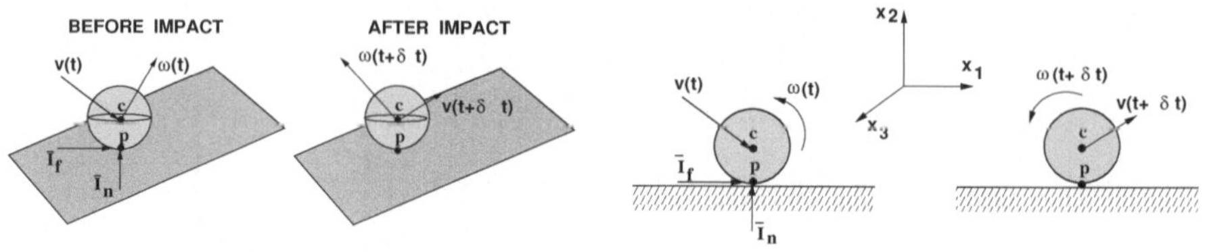

Fig. 4.8 A 3-D impact analysis and a 2-D schematic (Zohdi [4])

5. $\mathcal{M}_4^{Q_i-A}$ is the kurtosis (fourth moment), which measures the "tightness" of the distribution.

For our purposes here, the weight, $a_i = m_i$. This is straightforward to implement, and can provide much more detailed information on post-impact droplet characteristics.

4.3.3 Observation 2: Enhanced Kinematics-Rolling and Spin

The introduction of rolling and spin into the model is questionable for a small object, idealized by a particle that can undergo only translation. In closing, we indicate to the reader some of the effects that one could expect with models that allow rolling. In order to make things transparent, *initially,* we consider an isolated uncharged sphere impacting a surface.[2] Consider a 2D impact analysis as shown in Fig. 4.8. The impulse in the normal direction for the center of mass, located at point c, is governed by

$$mv_{c,n}(t) + \overline{I}_n \delta t = mv_{c,n}(t+\delta t) \Rightarrow \overline{I}_n = m\frac{\left(v_{c,n}(t+\delta t) - v_{c,n}(t)\right)}{\delta t}, \tag{4.25}$$

while the tangential impulse is governed by

$$mv_{ct}(t) + \overline{I}_f \delta t = mv_{ct}(t+\delta t) \Rightarrow \overline{I}_f = m\frac{(v_{ct}(t+\delta t) - v_{ct}(t))}{\delta t}. \tag{4.26}$$

The coefficient of restitution dictates, for point p and the surface:[3]

$$\mathcal{E} = \frac{v_{p,n}(t+\delta t) - v_{surf,n}(t+\delta t)}{v_{surf,n}(t) - v_{p,n}(t)}, \tag{4.27}$$

where $0 \leq \mathcal{E} \leq 1$. Assuming that the surface's velocity is zero, we obtain

$$v_{p,n}(t+\delta t) = -v_{p,n}(t)\mathcal{E}. \tag{4.28}$$

[2] Later, we consider the multiple particle, charged case.

[3] $\mathcal{E}$ is used here to avoid confusion with similar symbols that appear later in the presentation.

The velocity of the point p is

$$\boldsymbol{v}_p = \boldsymbol{v}_c + \boldsymbol{v}_{c\to p} = \boldsymbol{v}_c + \boldsymbol{\omega} \times \boldsymbol{r}_{c\to p} = \boldsymbol{v}_c + \omega R \boldsymbol{e}_t, \tag{4.29}$$

where R is the sphere radius and e_t is the unit vector in the tangential direction. The normal components are related by

$$v_{p,n} = \boldsymbol{v}_p \cdot \boldsymbol{e}_n = (\boldsymbol{v}_c + \omega R \boldsymbol{e}_t) \cdot \boldsymbol{e}_n = v_{c,n}, \tag{4.30}$$

where e_n is the unit vector in the normal direction, and thus $v_{p,n}(t+\delta t) = -v_{p,n}(t)\mathcal{E}$ implies

$$v_{c,n}(t+\delta t) = -v_{c,n}(t)\mathcal{E}, \tag{4.31}$$

and the tangential components by

$$v_{p,t} = \boldsymbol{v}_p \cdot \boldsymbol{e}_t = (\boldsymbol{v}_c + \omega R \boldsymbol{e}_t) \cdot \boldsymbol{e}_t = v_{ct} + \omega R \tag{4.32}$$

and thus

$$v_{ct}(t+\delta t) = v_{p,t}(t+\delta t) - \omega(t+\delta t) R, \tag{4.33}$$

where the normal force is

$$\overline{I}_n = -m \frac{v_{c,n}(t)(1+\mathcal{E})}{\delta t}, \tag{4.34}$$

and the friction force is

$$\overline{I}_f = m \left(\frac{v_{p,t}(t+\delta t) - \omega(t+\delta t) R - v_{ct}(t)}{\delta t} \right). \tag{4.35}$$

However, $\omega(t+\delta t)$ is unknown. Computing the angular momentum in the $\boldsymbol{e}_o = \boldsymbol{e}_z$ direction about point c yields $\omega(t+\delta t)$

$$\overline{\mathcal{I}}\omega(t) + \overline{I}_f R \delta t = \overline{\mathcal{I}}\omega(t+\delta t) \Rightarrow \omega(t+\delta t) = \omega(t) + \frac{\overline{I}_f R \delta t}{\overline{\mathcal{I}}}, \tag{4.36}$$

where $\overline{\mathcal{I}}$ is the moment of inertia with respect to the center of mass. However, $\overline{I}_f$ is a function of $\omega(t+\delta t)$ and obtains the following

$$\omega(t+\delta t) = \omega(t) + \frac{\left(m\left(v_{p,t}(t+\delta t) - \omega(t+\delta t) R - v_{ct}(t)\right)\right) R}{\overline{\mathcal{I}}}. \tag{4.37}$$

If we assume that there is *no slip*, then $v_{p,t}(t+\delta t) = 0$, and

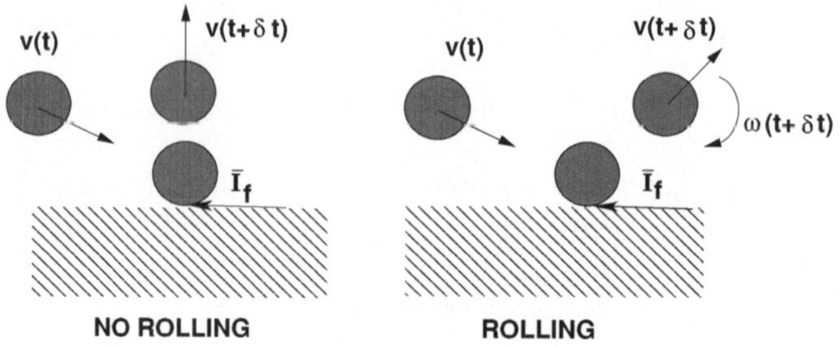

Fig. 4.9 A comparison of a nonrolling idealized particle (*left*) and a "particle" capable of spin (*right*) (Zohdi [4])

$$\omega(t+\delta t) = \omega(t) - \frac{mR\left(\omega(t+\delta t)R + v_{ct}(t)\right)}{\overline{\mathcal{I}}}, \tag{4.38}$$

and thus

$$\omega(t+\delta t) = \left(1 + \frac{mR^2}{\overline{\mathcal{I}}}\right)^{-1}\left(\omega(t) - \frac{mR}{\overline{\mathcal{I}}}v_{ct}(t)\right). \tag{4.39}$$

This relation is important. In the special case when $\omega(t) = 0$, one has, with $\overline{\mathcal{I}} = \frac{2}{5}mR^2$,

$$\omega(t+\delta t) = -\frac{5}{7R}v_{ct}(t), \tag{4.40}$$

and

$$v_{ct}(t+\delta t) = \frac{5}{7}v_{ct}(t), \tag{4.41}$$

and the friction force

$$\overline{I}_f = -\frac{2}{7}\frac{m}{\delta t}v_{ct}(t). \tag{4.42}$$

The implication is that there will be a tangential component of the velocity after impact, as opposed to a nonrotating particle, which would have no tangential velocity [although still a normal component, dictated by a coefficient of restitution (Fig. 4.9)]. *The main point is that allowing the particles to roll will allow them to more easily leave (roll off) the surface.*

Remark One could easily introduce rolling resistance forces into such model if deemed important. In our applications, it is unnecessary (see Johnson [1] for more details).

We remark that it is possible that the spin can be eliminated (or even reversed) when

$$\left(\omega(t) - \frac{mR}{\overline{\mathcal{I}}}v_{ct}(t)\right) = 0 \Rightarrow v_{ct}(t) = \frac{\omega(t)\overline{\mathcal{I}}}{mR} \tag{4.43}$$

Fig. 4.10 Two spinning particles that are about to make contact at point **c** (Zohdi [4])

It is essentially this relation that dictates the result of the impact, and asserts when the velocity of the center of mass can overcome the incoming spin.

Numerical implementation for multiple charged particles: As for the translation components of velocity, with a staggering scheme framework, the implementation of spin is straightforward. We assume that the objects are spheres. The governing equations for the balance of angular momentum about the centers of the particles are

$$\overline{\mathcal{I}}_i \cdot \dot{\boldsymbol{\omega}}_i = \sum_{j=1}^{K} \boldsymbol{M}_{ij} = \sum_{j=1}^{K} \boldsymbol{r}_{i \to c_j} \times \boldsymbol{\Psi}_{ij}^{fric}, \tag{4.44}$$

where the electrical and near-field forces do not come explicitly into the calculations since they act (an idealization) through the objects' centers. Of course, these forces implicitly affect the spin by dictating the relative approach velocities. Integrating this equation using the trapezoidal rule, we have (with $\overline{\mathcal{I}} = \overline{\mathcal{I}_s}\mathbf{1}$)

$$\frac{\boldsymbol{\omega}_i(t+\Delta t) - \boldsymbol{\omega}_i(t)}{\Delta t} = \frac{1}{\overline{\overline{\mathcal{I}}}_{si}} \int_t^{t+\delta t} \left(\sum_{j=1}^{K} \boldsymbol{M}_{ij} \right) dt, \tag{4.45}$$

which we approximate, consistent with our previous approach:

$$\boldsymbol{\omega}_i(t+\Delta t) = \boldsymbol{\omega}_i(t) + \frac{\delta t}{\overline{\overline{\mathcal{I}}}_{si}} \left(\sum_{j=1}^{K} \boldsymbol{M}_{ij} \right)_{|t*}, \tag{4.46}$$

where $t \leq t^* \leq t + \delta t$. Note that $\boldsymbol{\omega}_i(t+\Delta t) = \boldsymbol{\omega}_i(t+\delta t)$ in this formulation.

Frictional contact: The friction formula used before in the presence of external fields (due to the near-fields and other external electric fields), must now be applied to the point of contact, taking into account rotations in the relative velocities (Fig. 4.10). Surprisingly, the result is the same, as we see from the following analysis (where we initially assume stick at the contact point). For particle i (cm denotes center of mass)

$$m_i v_{it}^{cm}(t) - \overline{I}_f \delta t + \overline{\mathcal{E}}_{it} \delta t = m_i v_{it}^{cm}(t+\delta t), \tag{4.47}$$

and from the kinematics (c denotes the common point of contact between particles)

$$\boldsymbol{v}_i^c = \boldsymbol{v}_i^{cm} + \boldsymbol{\omega}_i \times \boldsymbol{r}_{i\to c}, \tag{4.48}$$

which yields

$$m_i v_{it}^{cm}(t) - \overline{I}_f \delta t + \overline{\mathcal{E}}_{it}\delta t = m_i \left(v_t^c(t+\delta t) - (\boldsymbol{\omega}_i(t+\delta t) \times \boldsymbol{r}_{i\to c}) \cdot \boldsymbol{\tau}_c \right), \tag{4.49}$$

where v_t^c is the common velocity of particles i and j in the tangential direction at the contact point, and $\boldsymbol{\tau}_c$ is the tangential direction of relative velocity and the contact point. Similarly, for the jth particle we have

$$m_j v_{jt}(t) + \overline{I}_f \delta t + \overline{\mathcal{E}}_{jt}\delta t = m_i \left(v_t^c(t+\delta t) - (\boldsymbol{\omega}_j(t+\delta t) \times \boldsymbol{r}_{j\to c}) \cdot \boldsymbol{\tau}_c \right). \tag{4.50}$$

There are two unknowns, $\overline{I}_f$ and v_t^c. The main quantity of interest is $\overline{I}_f$, which can be solved for

$$\overline{I}_f = \frac{\left(\frac{\overline{\mathcal{E}}_{it}}{m_i} - \frac{\overline{\mathcal{E}}_{jt}}{m_j}\right)\delta t + (v_{it}^{cm}(t) - v_{jt}^{cm}(t)) + \left((\boldsymbol{\omega}_i(t+\delta t) \times \boldsymbol{r}_{i\to c}) \cdot \boldsymbol{\tau}_c - (\boldsymbol{\omega}_j(t+\delta t) \times \boldsymbol{r}_{j\to c}) \cdot \boldsymbol{\tau}_c\right)}{\left(\frac{1}{m_i} + \frac{1}{m_j}\right)\delta t}. \tag{4.51}$$

This result contains extra $\boldsymbol{\omega}$-terms that are not present in the non-spinning particle case, which forces one to solve this simultaneously (iteratively) with Eq. 4.46. As before for the non-spinning case, consistent with stick-slip models of Coloumb friction, one first assumes no slip occurs. If

$$|\overline{I}_f| > \mu_s |\overline{I}_n|, \tag{4.52}$$

where $\mu_s \geq \mu_d$ is the coefficient of *static* friction, then slip must occur and a dynamic sliding friction model is used. If sliding occurs, the friction force is assumed to be proportional to the normal force and opposite to the direction of relative tangent motion, i.e.

$$\boldsymbol{\Psi}_i^{fric} \stackrel{\text{def}}{=} \mu_d ||\boldsymbol{\Psi}^{con}|| \frac{\boldsymbol{v}_{jt} - \boldsymbol{v}_{it}}{||\boldsymbol{v}_{jt} - \boldsymbol{v}_{it}||} = -\boldsymbol{\Psi}_j^{fric}. \tag{4.53}$$

References

1. Johnson, K. (1985). *Contact mechanics*. Cambridge: Cambridge University Press.
2. Martin, P. (2009). *Handbook of deposition technologies for films and coatings* (3rd ed.). Amsterdam: Elsevier.
3. Martin, P. (2011). *Introduction to surface engineering and functionally engineered materials*. Hoboken: Scrivener and Elsevier.
4. Zohdi, T. I. (submitted). Numerical simulation of the impact and deposition of charged particulated droplets. *Journal of Computational Physics*.

Chapter 5
An Introduction to Mechanistic Modeling of Swarms

The modeling of the dynamics of swarms has recently become a topic of interest in a number of scientific communities. Due to the similarities to the modeling of charged particulate dynamics, we provide a brief introduction.

It has long been recognized that interactive cooperative behavior within biological groups or "swarms" is advantageous in avoiding predators or, vice versa, in capturing prey. For example, one of the primary advantages of a swarm-like decentralized decision making structure is that there is no leader and thus the vulnerability of the swarm is substantially reduced. Furthermore, the decision making is relatively simple and rapid for each individual, however, the aggregate behavior of the swarm can be quite sophisticated.

The modeling of swarm-like behavior has biological research origins, dating back at least to Breder [5]. It is commonly accepted that a central characteristic of swarm-like behavior is the trade off between long-range interaction and short-range repulsion between individuals. Models describing clouds or swarms of particles, where their interaction is constructed from attractive and repulsive forces, dependent on the relative distance between individuals, are commonplace. For reviews, see Gazi and Passino [12], Bender and Fenton [2] or Kennedy and Eberhart [16]. The field is quite large and encompasses a wide variety of applications, for example, the behavior of flocks of birds, schools of fish, the flow of traffic and crowds of human beings, to name a few. Loosely speaking, swarm analyses are concerned with the complex aggregate behavior of groups of simple members which are frequently treated as particles (for example, in Zohdi [23]).

A central objective of this introduction is to provide basic mechanistic models and numerical solution strategies for the direct simulation of the motion of swarms that can be achieved within a relatively standard computing equipment. Although the approach taken in the present work is based on mechanical force interaction, it is important to mention that there exist a large number of what one can term as "rule-driven" swarms, whereby interaction is not governed by the principles of mechanics, but by proximal instructions, for example, "if fellow swarm member gets close to me, attempt to retreat as far as possible" or "follow the leader," or "stay in clusters," etc. Many species, for example, ant colonies (Bonabeau et al. [4]), exhibit foraging-

T. I. Zohdi, *Dynamics of Charged Particulate Systems*, SpringerBriefs in Applied Sciences and Technology, DOI: 10.1007/978-3-642-28519-6_5,

type behavior, in addition to the trail-laying-trail-following mechanism for finding food sources. During the search for food, they deposit a chemical substance, namely, a *pheromone*, which decays over time. Fellow foragers (swarm members) detect and follow paths with a high pheromone concentration, i.e., where the food source is highly concentrated. Although this type of model can be useful in some applications, it will not be discussed in this work. Recent broad overviews of the field can be found in Kennedy and Eberhart [16] and Bonabeau et al. [4]. For instance, Dorigo et al. [8] presented an optimization algorithm based on the foraging behavior of ants which basically used a computer adaptation of the pheromone trail-laying-trail-following method to mimic the behavior of ants to allow the "software ants" to solve combinatorial problems such as the traveling salesman problem.[1] Bonabeau et al. [4] presented several such optimization algorithms, each one influenced by another feature of biological swarms. There are numerous other models in this direction that develop optimization techniques [4, 16], business planning [3], telecommunication network design [4], mobile sensor networks [11], robotics and vehicle navigation and military applications. Early approaches that rely on decentralized organization can be found in Brooks [6] and references therein. The related field of cooperative robotics is quite large; references and attempts to overview and classify the numerous publications can be found in Dudek et al. [9], Cao et al. [7], and Liu and Passino [17]. Broader overviews on the topic of swarm intelligence are given in Bonabeau et al. [4], and Kennedy and Eberhardt [16]. In the references, an extensive list of works has been included. While these rule-driven paradigms are usually easy to construct, they are difficult to analyze mathematically. It is primarily for this reason that a mechanical approach is adopted here, and follow the framework found in Zohdi [23].

5.1 A Basic Construction of a Swarm

In the following analysis, we treat the swarm members as point masses, i.e., we ignore their dimensions.[2] For each swarm member (N_s in total), the equations of motion are

$$m_i \ddot{\boldsymbol{r}}_i = \boldsymbol{\Psi}_i(\boldsymbol{r}_1, \boldsymbol{r}_2, ..., \boldsymbol{r}_{N_s}), \tag{5.1}$$

where $\boldsymbol{\Psi}_i$ represents the forces of interaction between swarm member i and the target, obstacles, and other swarm members. We consider the following decomposition of interaction forces

$$\boldsymbol{\Psi}_i = \boldsymbol{\Psi}_i^{mm} + \boldsymbol{\Psi}_i^{mt} + \boldsymbol{\Psi}_i^{mo}, \tag{5.2}$$

[1] Finding the least expensive route for traveling to a number of given locations, given the costs for each connection route.

[2] The swarm member centers, which are initially nonintersecting, cannot intersect later due to the singular repulsion terms.

where between swarm members (member-member)

$$\boldsymbol{\Psi}_i^{mm} = \sum_{j \neq i}^{N_s} \left(\left(\underbrace{\alpha_1^{mm} ||\boldsymbol{r}_i - \boldsymbol{r}_j||^{\beta_1^{mm}}}_{\text{attraction}} - \underbrace{\alpha_2^{mm} ||\boldsymbol{r}_i - \boldsymbol{r}_j||^{-\beta_2^{mm}}}_{\text{repulsion}} \right) \underbrace{\frac{\boldsymbol{r}_j - \boldsymbol{r}_i}{||\boldsymbol{r}_i - \boldsymbol{r}_j||}}_{\boldsymbol{n}_{ij} \stackrel{\text{def}}{=} \text{unit vector}} \right), \tag{5.3}$$

where $|| \cdot ||$ represents the Euclidean norm in R^3, and the normal direction is determined by the difference in the position vectors of the particles' centers

$$\boldsymbol{n}_{ij} \stackrel{\text{def}}{=} \frac{\boldsymbol{r}_j - \boldsymbol{r}_i}{||\boldsymbol{r}_i - \boldsymbol{r}_j||}. \tag{5.4}$$

Between the swarm members and the target, we have (member-target)

$$\boldsymbol{\Psi}_i^{mt} = \left(\alpha^{mt} ||\boldsymbol{r}_i - \boldsymbol{T}||^{\beta^{mt}} \right) \frac{\boldsymbol{T} - \boldsymbol{r}_i}{||\boldsymbol{r}_i - \boldsymbol{T}||}, \tag{5.5}$$

and for the repulsion between swarm members and the obstacles (member-obstacle)

$$\boldsymbol{\Psi}_i^{mo} = -\sum_{j=1}^{q} \left(\left(\alpha^{mo} ||\boldsymbol{r}_i - \boldsymbol{O}_j||^{-\beta^{mo}} \right) \frac{\boldsymbol{O}_j - \boldsymbol{r}_i}{||\boldsymbol{r}_i - \boldsymbol{O}_j||} \right), \tag{5.6}$$

where q is the number of obstacles and where all of the (design) parameters, α's and β's, are nonnegative.

Remark 1 One can describe the relative contributions of repulsion and attraction between members of the swarm by considering an individual pair in (static) equilibrium

$$\boldsymbol{\Psi}^{mm} = \left(\alpha_1^{mm} ||\boldsymbol{r}_i - \boldsymbol{r}_j||^{\beta_1^{mm}} - \alpha_2^{mm} ||\boldsymbol{r}_i - \boldsymbol{r}_j||^{-\beta_2^{mm}} \right) \frac{\boldsymbol{r}_j - \boldsymbol{r}_i}{||\boldsymbol{r}_i - \boldsymbol{r}_j||} = \boldsymbol{0}. \tag{5.7}$$

This characterizes a separation length scale describing the tendency to cluster or spread apart, that is,

$$||\boldsymbol{r}_i - \boldsymbol{r}_j|| = \left(\frac{\alpha_2^{mm}}{\alpha_1^{mm}} \right)^{\frac{1}{\beta_1^{mm} + \beta_2^{mm}}} \stackrel{\text{def}}{=} d_e^{mm}. \tag{5.8}$$

If the distance by which the swarm members can communicate is denoted d^{com}, and $d^{com} \leq d_e^{mm}$, then there will possibly be no interaction, for example, at static equilibrium.

Remark 2 The specific structure of the interparticle forces chosen is only one of many possibilities to model the interaction. There are numerous other possibilities. The properties of this specific type of representation, such as, the work expenditure, energy and power, are discussed in the Appendix. There are a variety of alternative forms available from the field of Molecular Dynamics (MD), which is typically concerned with the calculation of thermochemical and thermomechanical properties of gases, liquids and solids by using models of systems of atoms or molecules where each atom (or molecule) is represented by a material point and is treated as a point mass whose motion is described by the Newton's second law with the forces computed from a prescribed potential energy function, $V(\boldsymbol{r})$, $m\ddot{\boldsymbol{r}} = -\nabla V(\boldsymbol{r})$ (see Haile [13], for example).

5.1.1 Environmental Damping

A source of damping for the system is from the (surrounding) environment (for example, a fluid such as air). The simplest model is of the form (for swarm member i)

$$\boldsymbol{\Psi}_i^{env} = -c^{env}(\boldsymbol{v}_i - \boldsymbol{v}^{env}) \tag{5.9}$$

where $\boldsymbol{v}_i$ is the velocity of the ith member and $\boldsymbol{v}^{env}$ is the local velocity of the ambient medium. In summary, we have the following forces acting on each member of the swarm, that is,

$$\boldsymbol{\Psi} = \boldsymbol{\Psi}^{mm} + \boldsymbol{\Psi}^{mt} + \boldsymbol{\Psi}^{mo} + \boldsymbol{\Psi}^{env}. \tag{5.10}$$

Remark The problem of fully coupled (two-way) particle-fluid interaction is beyond the scope of the present work. Generally, this requires the use of staggering-type schemes [22]. The numerical simulation of the motion of the swarm members utilizes the same algorithms that have been introduced earlier in the monograph for charged particles, however, without contact and without thermal effects. An implementation of the procedure is as follows:

(1) GLOBAL FIXED – POINT ITERATION : (SET i = 1 AND K = 0) :
(2) IF i > $\mathrm{N_p}$ THEN GO TO (4)
(3) IF i ≤ $\mathrm{N_p}$ THEN :
 (*a*) COMPUTE POSITION : $\boldsymbol{r}_i^{L+1,K}$
 (*b*) GO TO (2) FOR NEXT SWARM MEMBER (i = i + 1)
(4) ERROR MEASURE :

$$(a)(\mathrm{ERROR})_K \stackrel{\mathrm{def}}{=} \frac{\sum_{i=1}^{N_s} ||\boldsymbol{r}_i^{L+1,K} - \boldsymbol{r}_i^{L+1,K-1}||}{\sum_{i=1}^{N_s} ||\boldsymbol{r}_i^{L+1,K} - \boldsymbol{r}_i^{L}||} \quad \text{(normalized)}$$

$$(b) Z_K \stackrel{\mathrm{def}}{=} \frac{(\mathrm{ERROR})_K}{TOL_r}$$

$$(c)\Phi_K \stackrel{\mathrm{def}}{=} \left(\frac{\left(\frac{TOL}{(\mathrm{ERROR})_0}\right)^{\frac{1}{pK_d}}}{\left(\frac{(\mathrm{ERROR})_K}{(\mathrm{ERROR})_0}\right)^{\frac{1}{pK}}} \right)$$

(5) IF TOLERANCE NOT MET ($Z_K > 1$) AND $K < K_d$ REPEAT ITERATION (K = K + 1)
(6) IF TOLERANCE MET ($Z_K \leq 1$) AND $K < K_d$ THEN :
  (a) INCREMENT TIME : $t = t + \Delta t$
  (b) CONSTRUCT NEW TIME STEP : $\Delta t = \Phi_K \Delta t$,
  (c) SELECT MINIMUM : $\Delta t = MIN(\Delta t^{lim}, \Delta t)$ AND GO TO (1)
(7) IF TOLERANCE NOT MET ($Z_K > 1$) AND $K = K_d$ THEN :
  (a) CONSTRUCT NEW TIME STEP : $\Delta t = \Phi_K \Delta t$
  (b) RESTART AT TIME = t AND GO TO (1).

(5.11)

5.2 Numerical Examples

5.2.1 A Model Problem: Chasing a Moving Target

As a representative of a class of model problems, we now consider a normalized "performance" function (normalized by the total simulation time and the initial separation distance) representing (1) the time it takes for the swarm members to get to the target and (2) the distance of the swarm members away from the target:

$$\Pi = \frac{\left(\int_0^{\mathcal{T}} \sum_{i=1}^{N_p} ||\boldsymbol{r}_i - \boldsymbol{T}||\, dt\right)}{\mathcal{T} \sum_{i=1}^{N_p} ||\boldsymbol{r}_i(t=0) - \boldsymbol{T}||}, \tag{5.12}$$

where total simulation time is $\mathcal{T} = 30$ s and $\boldsymbol{T}$ is the position of the target (Fig. 5.1). The components of the initial position vectors of the nonintersecting swarm members (each assigned a mass of 10 kg) were given random values of $-1 \leq r_{ix}, r_{iy}, r_{iz} \leq 1$. The location of the *moving* target (parameters in Table. 5.1) was given by

$$\begin{aligned} T_x &= x_o + a_1 cos(a_2 t) + a_3 t, \\ T_y &= y_o + b_1 sin(b_2 t) + b_3 t, \\ T_z &= z_o + c_1 cos(c_2 t) + c_3 t. \end{aligned} \tag{5.13}$$

The location of the center of the (rectangular) obstacle array was (1.5, 0, 0). A 100-obstacle "fence" was set up in a 10×10 array with a spacing of 0.2 between obstacle centers. For illustrative purposes, 200 swarm members were used. The parameters selected were: $\alpha_1^{mm} = 1$, $\alpha_2^{mm} = 1$, $\alpha^{mt} = 200$, $\alpha^{mo} = 100$, $\beta_1^{mm} = 2$, $\beta_2^{mm} = 2$, $\beta^{mt} = 2$ and $\beta^{mo} = 2$. The environmental damping was set to $c^{env} = 1$. The total time was set to $\mathcal{T} = 30$. Simulations, shown in Fig. 5.2, were run with the performance being $\Pi = 0.2712$ (time steps: 3439, fixed-point iterations: 24633). Although there is environmental damping, the envelope of motion is initially quite large (Figure 5.2).

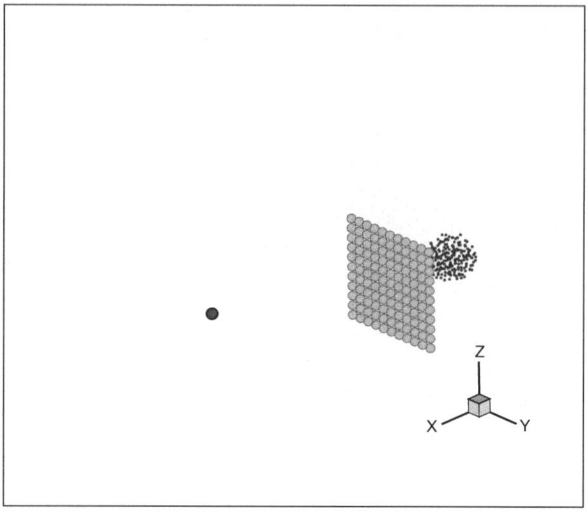

Fig. 5.1 A typical arrangement for a "swarm." The target is shown in *red*, the obstacles in *green* and the swarm members are shown in *blue* (Zohdi [23])

Remark 1 Typically, for systems with a finite number of particles, there will be slight variations in the performance for different random starting configurations. In order to stabilize the objective function's value with respect to the randomness of the flow starting configuration, for a given parameter selection ($\boldsymbol{\Lambda}$, characterized by the α's and β's), a regularization procedure is applied, whereby the performances of a series of different random starting configurations are averaged until the (ensemble) average converges, i.e., until the following condition is met:

$$\left|\frac{1}{E+1}\sum_{i=1}^{E+1}\Pi^{(i)}(\boldsymbol{\Lambda}^I)-\frac{1}{E}\sum_{i=1}^{E}\Pi^{(i)}(\boldsymbol{\Lambda}^I)\right| \leq TOL\left|\frac{1}{E+1}\sum_{i=1}^{E+1}\Pi^{(i)}(\boldsymbol{\Lambda}^I)\right|, \tag{5.14}$$

where index i indicates a different starting random configuration ($i = 1, 2, \ldots, E$) that has been generated and E indicates the total number of configurations tested. For swarms of the sizes tested, typically, two or three sample realizations were needed for a stable ensemble average.

Remark 2 In Zohdi [21], different sized swarms were tested, and the resulting optimal strategies (attraction and repulsion coefficients) were tabulated. From those results, it became clear that in some cases, if the swarm is small enough, bunching up and moving through the obstacle course is the optimal strategy. Generally, the best strategy strongly depends on the obstacle course size and shape, and swarm size. A strategy for estimating the parameters, based on genetic algorithms, is given in Zohdi [21].

Remark 3 If a swarm member gets too close to an obstacle, we may want to enforce that it becomes immobilized. In this case, a side condition can be introduced using

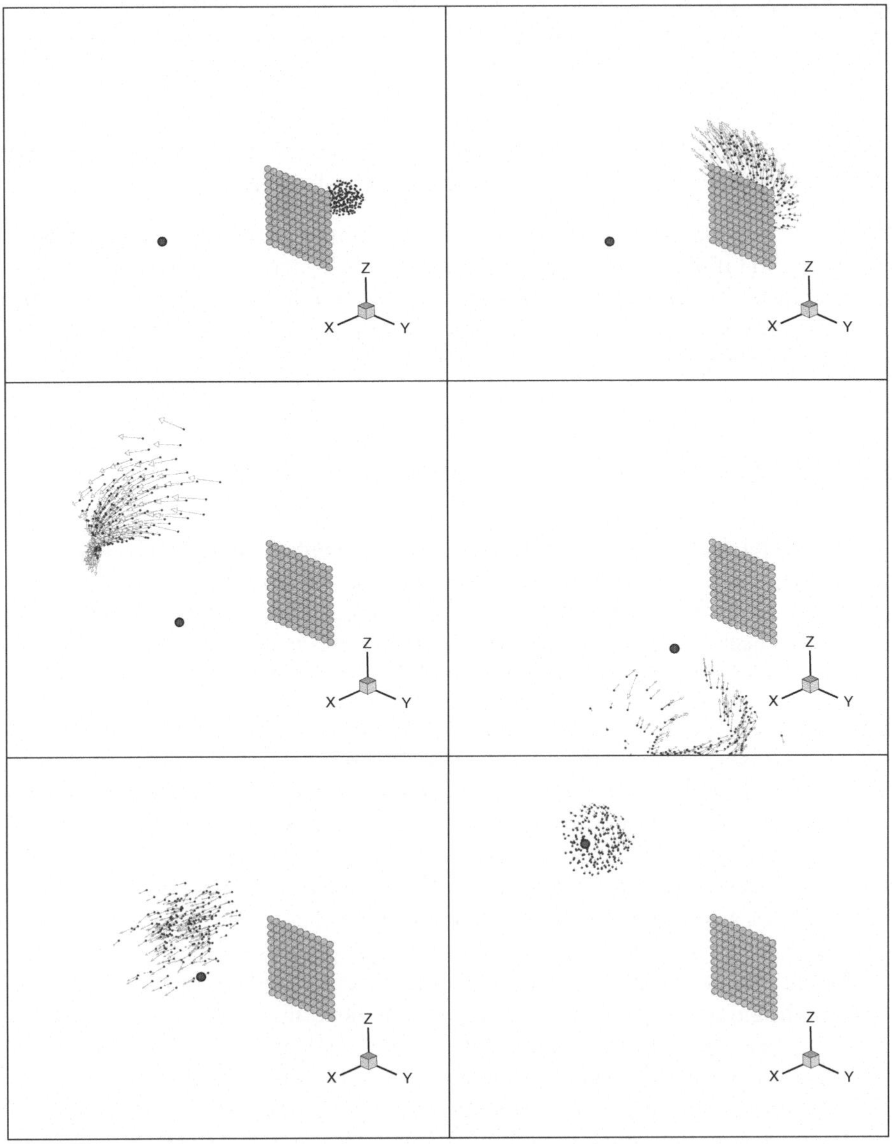

Fig. 5.2 *Top* to *bottom* and *left* to *right*: the swarm moves over the obstacle fence (Zohdi [23])

the form $\forall\, t$, $\forall\, r_{oj}$ and $\tau < \mathcal{T}$, if

$$||\boldsymbol{r}_i(t = \tau) - \boldsymbol{O}_j|| \leq R. \tag{5.15}$$

Then, $\boldsymbol{r}_i = \boldsymbol{r}_i(t = \tau)$, $\forall\, t \geq \tau$, where the unilateral condition represents the effect of being near a "destructive" obstacle. The swarm member is stopped in the position

Table 5.1 Table of parameters

(x_o, y_o, z_o)	a_1	a_2	a_3	b_1	b_2	b_3	c_1	c_2	c_3
(4,0,0)	1	1	0.5	1	1	0.5	1	1	0.5

where it enters the "radius of destruction" (R). Therefore, the swarm performance (Π) is severely penalized if it loses members to the obstacles.

Remark 4 It is important to note that if the interaction is only between the nearest neighbors, and if there is no inertial reference point for the swarm members to refer to, instabilities (collisions) may occur [14, 15, 18–20]. In the present analysis, such inertial reference points were provided by the swarm's knowledge of the stationary obstacles and target.

5.2.2 Another Model Problem: Multisite Search

As another model problem, consider 400 swarm members and 200 randomly dispersed "target sites" which the swarm is tasked to visit (Fig. 5.3). The algorithm is as follows: (1) Each swarm member is attracted to the nearest target location and (2) If a site has been visited, then it is inactive (the swarm is not attracted to it). As the frames indicate, the swarm has a natural tendency to divide and conquer the domain.

Remark There are over 100,000,000,000 websites as of 2007. There are, on average, approximately 250 words per page (like a book). Clearly, searches for a piece of information would take a very long time if done directly. Since a computer cannot recognize words and sentences, only bytes, this requires specialized natural language processing algorithms which incorporate programs for parsing. In this regard, there are processes such as "web-crawling" or "spidering." These approaches systematically browse the net in an automated fashion. These programs are primarily used to create a copy of all visited pages for later processing by a search engine. It is hoped that swarm-type search, if adapted properly, and combined with proper web-crawling routines, could provide a new, and hopefully faster search paradigm. Primarily, here, we mean by search (both virtual and physical): (1) *Data mining* for informational queries (via the Internet), (2) *Multipronged cloud-type search:* circumvents the slower tree-type data structure search and (3) *Area-coverage/man-overboard search:* determination of optimal paths for maximum area-coverage, for example, for a lost object.

5.3 Discussion and Concluding Remarks

In many applications, the computed positions, velocities and accelerations of the members of a swarm, for example, people or vehicles, must be translated into realizable movement. Furthermore, the communication latency and information exchange

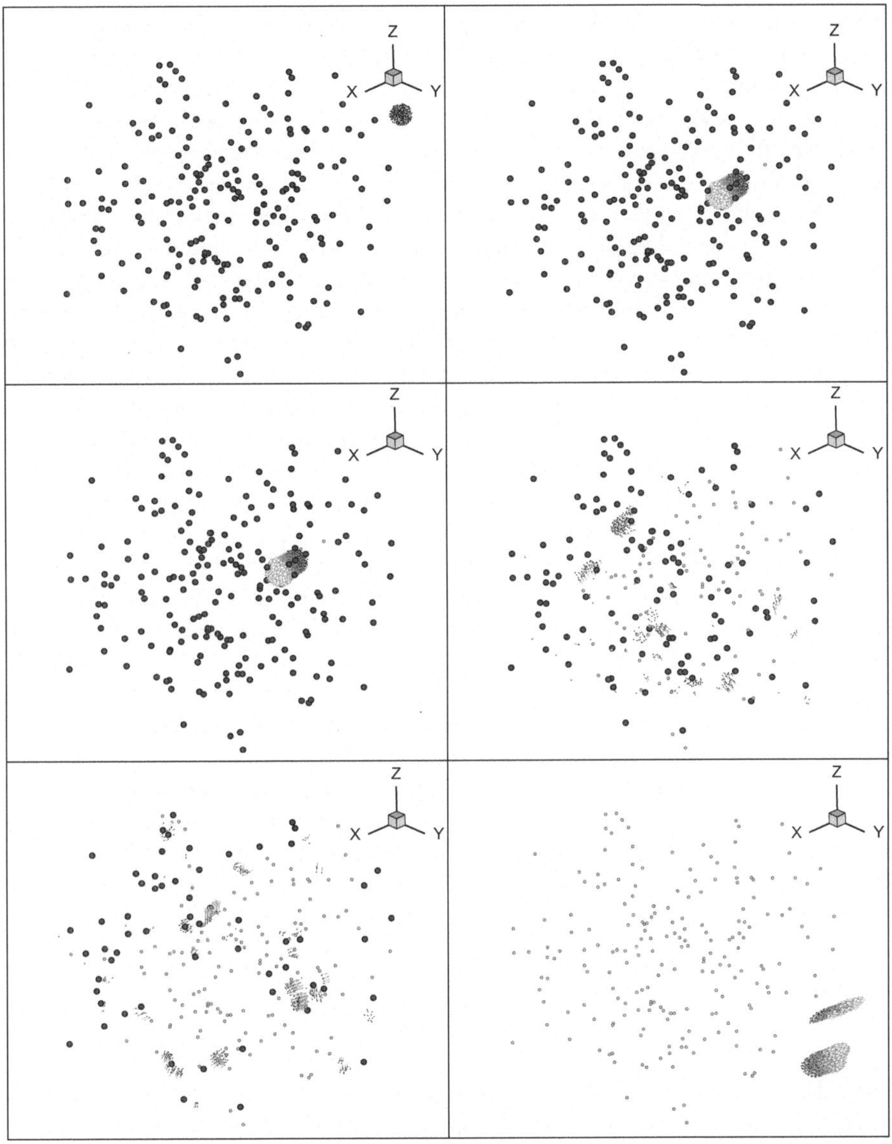

Fig. 5.3 *Top* to *bottom* and *left* to *right*: the swarm moves through the search space. *Red* represents sites unvisited and *green* represents sites visited (Zohdi [23])

poses a significant technological hurdle. In practice, further sophistication, i.e., constraints on movement and communication, must be embedded into the computational model for the application at hand. However, the fundamental computational philosophy and modeling strategy should remain relatively unchanged.

For certain types of swarms, the "visual field" of the individual members may play a significant role in the overall behavior. In some cases, this is a nonissue, for example, if the vehicles are robots or UAVs, since the communication is most likely electronic. However, certain animals see only directly ahead of them. A relatively simple way to incorporate this into a simulation is to check the inner product of each swarm member's velocity with a neighbor's relative position vector, $(\boldsymbol{r}_j - \boldsymbol{r}_i) \cdot \boldsymbol{v}_i$. Under the assumption that the swarm members "look where they are going," if the inner product is negative, this indicates that the neighbor is behind the swarm member's visual field, and hence there is no interaction between this specific pair of swarm members. Also, it is important to note that some groups interact with their nearest neighbors, while some do so with a specific number of swarm members, *regardless* of whether they are far away [10]. For example, specifically for Starlings (Sturnus vulgaris), Ballerini et al. [1] conclude, based on a number of careful observations, that interactions are governed by topological distance and not metric distance, i.e., a bird communicates with a certain number of birds surrounding it, *regardless of the distance away.* Ballerini et al. [1] believe that this may be attributed to a perceptual limit in the number of objects that they can track. Specifically, Ballerini et al. [1] have determined that the interaction, for a typical bird in a swarm, decays rapidly after approximately the sixth or seventh nearest neighbor. For Starlings, their empirical evidence appears to support this hypothesis. This issue is currently under investigation by the author.

Finally, an important aspect of any model is the identification of parameters which force the system behavior to approximate, as close as possible, a desired target response. For example, in the ideal case, one would like to determine the type of interaction that produces certain overall system characteristics, via numerical simulations, in order to guide or minimize time-consuming laboratory tests. As a representative of a class of model problems, consider "inverse" problems whereby the parameters in the interaction representation are sought, the α's and β's, which deliver a target system behavior by minimizing a normalized cost function

$$\Pi = \frac{\int_0^{\mathcal{T}} |A - A^*| \, dt}{\int_0^{\mathcal{T}} |A^*| \, dt}, \tag{5.16}$$

where the total simulation time is $\mathcal{T}$, where A is a computationally generated quantity of interest and where A^* is the target response. Typically, for the class of problems considered in this work, formulations (Π), for example, in Eq. 5.16, depend, in a nonconvex and nondifferentiable manner, on the α's and β's. This is primarily due to the nonlinear character of the interaction, the physics of sudden interparticle impact and the transient dynamics. Clearly, we must have restrictions (for physical reasons) on the parameters in the interaction

$$\alpha_i^- \le \alpha_i \le \alpha_i^+ \tag{5.17}$$

and

$$\beta_i^- \leq \beta_i \leq \beta_i^+, \tag{5.18}$$

where α_i^-, α_i^+, β_i^- and β_i^+, are the lower and upper limits on the coefficients in the interaction forces. Typically, the class of the objective functions represented by Eq. 5.16 are nonconvex and nondifferentiable, and similar in structure to the inverse particulate flow problems discussed in the previous chapter. We refer the reader to that section of the text for comments on minimization strategies.

References

1. Ballerini, M., Cabibbo, N., Candelier, R., Cavagna, A., Cisbani, E., Giardina, I., et al. (2008). Interaction ruling animal collective behavior depends on topological rather than metric distance: Evidence from a field study. *PNAS, 105*(4), 1232–1237.
2. Bender, J., & Fenton, R. (1970). On the flow capacity of automated highways. *Transport Science, 4*, 52–63.
3. Bonabeau, E., & Meyer, C. (2001). Swarm intelligence: A whole new way to think about business. *Harvard Business Review, 79*(5), 106–114.
4. Bonabeau, E., Dorigo, M., & Theraulaz, G. (1999). *Swarm intelligence: From natural to artificial systems*. New York: Oxford University Press.
5. Breder, C. M. (1954). Equations descriptive of fish schools and other animal aggregations. *Ecology, 35*(3), 361–370.
6. Brooks, R. A. (1991). Intelligence without reason. *Proceedings of the 12th International Joint Conference on Artificial Intelligence* (IJCAI-91), (pp. 569–595).
7. Cao, Y. U., Fukunaga, A. S., & Kahng, A. (1997). Cooperative mobile robotics: Antecedents and directions. *Autonomous Robots, 4*(1), 7–27.
8. Dorigo, M., Maniezzo, V., & Colorni, A. (1996). Ant system: optimization by a colony of cooperating agents. *IEEE Transactions on Systems, Man and Cybernetics, Part B, 26*(1), 29–41.
9. Dudek, G., Jenkin, M., Milios, E., & Wilkes, D. (1996). A taxonomy for multi-agent robotics. *Autonomous Robots, 3*, 375–397.
10. Feder, T. (2007). Statistical physics is for the birds. *Physics Today, 60*, 28–29.
11. Fiorelli, E., Leonard, N. E., Bhatta, P., Paley, D., Bachmayer, R., & Fratantoni, D. M. (2004). Multi-AUV control and adaptive sampling in Monterey Bay. In Autonomous Underwater Vehicles, 2004 IEEE/OES (pp. 134–147).
12. Gazi, V., & Passino, K. M. (2002). Stability analysis of swarms. *Proceedings of the American Control Conference,* Anchorage, AK, May 8–10.
13. Haile, J. M. (1992). *Molecular dynamics simulations: Elementary methods*. Chichester: Wiley.
14. Hedrick, J. K., & Swaroop, D. (1993). Dynamic coupling in vehicles under automatic control. In *13th IAVSD Symposium,* August (pp. 209–220).
15. Hedrick, J. K., Tomizuka, M., & Varaiya, P. (1994). Control issues in automated highway systems. *IEEE Control Systems Magazine, 14*(6), 21–32.
16. Kennedy, J., & Eberhart, R. (2001). *Swarm Intelligence*. San Francisco: Morgan Kaufmann Publishers.
17. Liu, Y., & Passino, K. M. (2000) *Swarm intelligence: Literature overview.* Technical report, Ohio State University.
18. Shamma, J. S. (2001). A connection between structured uncertainty and decentralized control of spatially invariant systems. *Proceedings of the 2001 American Control Conference,* Arlington, VA. Part vol. 4 (pp. 3117–3121).

19. Swaroop, D., & Hedrick, J. K. (1996). String stability of interconnected systems. *IEEE Transactions on Automatic Control, 41*(4), 349–356.
20. Swaroop, D., & Hedrick, J. K. (1999). Constant spacing strategies for platooning in automated highway systems. *Journal of Dynamic Systems, Measurement, and Control Transaction, 121*, 462–470.
21. Zohdi, T. I. (2003). Computational design of swarms. *International Journal for Numerical Methods in Engineering, 57*, 2205–2219.
22. Zohdi, T. I. (2007). Computation of strongly coupled multifield interaction in particle-fluid systems. *Computer Methods in Applied Mechanics and Engineering, 196*, 3927–3950.
23. Zohdi, T. I. (2009). Mechanistic modeling of swarms. *Computer Methods in Applied Mechanics and Engineering, 198*(21–26), 2039–2051.

Appendix
Construction of Potentials

A.1 Work, Energy and Power

The differential amount of work done by a force acting through a differential displacement is

$$dW = \Psi \cdot d\mathbf{r}. \tag{A.1}$$

Therefore, the total amount of work performed by a force over a displacement history is

$$W_{1\to 2} = \int_{\boldsymbol{r}(t_1)}^{\boldsymbol{r}(t_2)} \Psi \cdot d\boldsymbol{r} = \int_{\boldsymbol{r}(t_1)}^{\boldsymbol{r}(t_2)} m\boldsymbol{a} \cdot d\boldsymbol{r} = \int_{\boldsymbol{v}(t_1)}^{\boldsymbol{v}(t_2)} m\boldsymbol{v} \cdot d\boldsymbol{v}$$
$$= \frac{1}{2} m(\boldsymbol{v}_2 \cdot \boldsymbol{v}_2 - \boldsymbol{v}_1 \cdot \boldsymbol{v}_1) \stackrel{\text{def}}{=} T_2 - T_1, \tag{A.2}$$

where $T \stackrel{\text{def}}{=} \frac{1}{2} m\boldsymbol{v} \cdot \boldsymbol{v}$ is known as the kinetic energy. The chain rule was used to write $\boldsymbol{a} \cdot d\boldsymbol{r} = \boldsymbol{v} \cdot d\boldsymbol{v}$. Therefore, we may write

$$T_1 + W_{1\to 2} = T_2. \tag{A.3}$$

If the forces can be written in the following form

$$dV = -\Psi \cdot d\boldsymbol{r}, \tag{A.4}$$

where

$$\Psi = -\nabla V, \tag{A.5}$$

then

$$W_{1\to 2} = -\int_{\boldsymbol{r}(t_1)}^{\boldsymbol{r}(t_2)} dV = V(\boldsymbol{r}(t_1)) - V(\boldsymbol{r}(t_2)). \tag{A.6}$$

T. I. Zohdi, *Dynamics of Charged Particulate Systems*, SpringerBriefs in Applied Sciences and Technology, DOI: 10.1007/978-3-642-28519-6,

Such a force is said to be conservative. Furthermore, it is easy to show that a conservative force must satisfy

$$\nabla \times \boldsymbol{\Psi} = \mathbf{0}. \tag{A.7}$$

The work done by a conservative force on any closed path is zero since

$$-\int_{\boldsymbol{r}(t_1)}^{\boldsymbol{r}(t_2)} dV = V(\boldsymbol{r}(t_1)) - V(\boldsymbol{r}(t_2)) = \int_{\boldsymbol{r}(t_2)}^{\boldsymbol{r}(t_1)} dV \Rightarrow \int_{\boldsymbol{r}(t_1)}^{\boldsymbol{r}(t_2)} dV + \int_{\boldsymbol{r}(t_2)}^{\boldsymbol{r}(t_1)} dV = 0. \tag{A.8}$$

As a consequence, for a conservative system,

$$T_1 + V_1 = T_2 + V_2. \tag{A.9}$$

Also, power can be defined as the time rate of change of work

$$\frac{dW}{dt} = \frac{\boldsymbol{\Psi} \cdot d\boldsymbol{r}}{dt} = \boldsymbol{\Psi} \cdot \boldsymbol{v}. \tag{A.10}$$

A.2 Properties of a Potential

As we have indicated, a force field $\boldsymbol{\Psi}$ is said to be conservative if and only if there exists a continuously differentiable scalar field V such that $\boldsymbol{\Psi} = -\nabla V$. Therefore, a necessary and sufficient condition for a particle to be in equilibrium is that

$$\boldsymbol{\Psi} = -\nabla V = \mathbf{0}. \tag{A.11}$$

In other words,

$$\frac{\partial V}{\partial x_1} = 0, \ \frac{\partial V}{\partial x_2} = 0 \text{ and } \frac{\partial V}{\partial x_3} = 0. \tag{A.12}$$

Forces acting on a particle that are (1) always directed toward or away from another point and (2) whose magnitude only depends on the distance between the particle and the point of attraction/repulsion are called *central forces*. They have the form

$$\boldsymbol{\Psi} = -\mathcal{C}(||\boldsymbol{r} - \boldsymbol{r}_o||)\frac{\boldsymbol{r} - \boldsymbol{r}_o}{||\boldsymbol{r} - \boldsymbol{r}_o||} = \mathcal{C}(||\boldsymbol{r} - \boldsymbol{r}_o||)\boldsymbol{n}, \tag{A.13}$$

where $\boldsymbol{r}$ is the position of the particle, $\boldsymbol{r}_o$ is the position of a point to which the particle is attracted toward and repulsed from and

$$\boldsymbol{n} = \frac{\boldsymbol{r}_o - \boldsymbol{r}}{||\boldsymbol{r} - \boldsymbol{r}_o||}. \tag{A.14}$$

The central force is one of attraction if

$$\mathcal{C}(||\boldsymbol{r} - \boldsymbol{r}_o||) > 0 \tag{A.15}$$

and one of repulsion if

$$\mathcal{C}(||\boldsymbol{r} - \boldsymbol{r}_o||) < 0. \tag{A.16}$$

We remark that a central force field is always conservative since $\nabla \times \Psi = \mathbf{0}$. Now, consider the specific choice

$$V = \underbrace{\frac{\alpha_1 ||\boldsymbol{r} - \boldsymbol{r}_o||^{-\beta_1+1}}{-\beta_1 + 1}}_{\text{attraction}} - \underbrace{\frac{\alpha_2 ||\boldsymbol{r} - \boldsymbol{r}_o||^{-\beta_2+1}}{-\beta_2 + 1}}_{\text{repulsion}} \tag{A.17}$$

where all of the parameters, α's and β's are nonnegative. The gradient yields

$$-\nabla V = \Psi = \left(\alpha_1 ||\boldsymbol{r} - \boldsymbol{r}_o||^{-\beta_1} - \alpha_2 ||\boldsymbol{r} - \boldsymbol{r}_o||^{-\beta_2}\right)\boldsymbol{n}, \tag{A.18}$$

which is repeatedly used in this monograph. If a particle which is displaced slightly from an equilibrium point tends to return to that point, then we call that point a point of stability or a stable point, and the equilibrium is said to be stable. Otherwise, we say that the point is one of instability and the equilibrium is unstable. *A necessary and sufficient condition for a point of equilibrium to be stable* is that the potential V at that point be a minimum. The general condition by which a potential is stable for the multidimensional case can be determined by studying the properties of the Hessian of V,

$$[\boldsymbol{I\!H}] \stackrel{\text{def}}{=} \begin{bmatrix} \frac{\partial^2 V}{\partial x_1 \partial x_1} & \frac{\partial^2 V}{\partial x_1 \partial x_2} & \frac{\partial^2 V}{\partial x_1 \partial x_3} \\ \frac{\partial^2 V}{\partial x_2 \partial x_1} & \frac{\partial^2 V}{\partial x_2 \partial x_2} & \frac{\partial^2 V}{\partial x_2 \partial x_3} \\ \frac{\partial^2 V}{\partial x_3 \partial x_1} & \frac{\partial^2 V}{\partial x_3 \partial x_2} & \frac{\partial^2 V}{\partial x_3 \partial x_3} \end{bmatrix}, \tag{A.19}$$

around an equilibrium point. A sufficient condition for V to attain a minimum at an equilibrium point is for the Hessian to be positive definite (which implies that V is locally convex). For more details, see Hale and Kocak [67].

Remark Provided that the α's and β's are appropriately selected, the chosen central force potential form is stable for motion in the normal direction, i.e., the line connecting the centers of particles in particle-particle interaction. For disturbances in directions orthogonal to the normal direction, the potential is neutrally stable, i.e., the Hessian's determinant is zero, thus indicating that the potential does not change for such perturbations. The motion analysis in the normal direction is relevant for central forces of the type under consideration. In order to determine stable parameter combinations, consider a potential function for a single particle, in one-dimensional motion, representing the motion in the normal direction, attracted and repulsed from a point r_o, measured by the coordinate r,

$$V = \frac{\alpha_1}{-\beta_1 + 1}|r - r_o|^{-\beta_1+1} - \frac{\alpha_2}{-\beta_2 + 1}|r - r_o|^{-\beta_2+1}, \tag{A.20}$$

whose derivative produces the form of interaction forces introduced earlier:

$$\Psi = -\frac{\partial V}{\partial r} = \left(\alpha_1|r - r_o|^{-\beta_1} - \alpha_2|r - r_o|^{-\beta_2}\right)n, \tag{A.21}$$

where $n = \frac{r_o - r}{|r - r_o|}$. For stability, we require

$$\frac{\partial^2 V}{\partial r^2} = -\alpha_1\beta_1|r - r_o|^{-\beta_1-1} + \alpha_2\beta_2|r - r_o|^{-\beta_2-1} > 0. \tag{A.22}$$

A static equilibrium point, $r = r_e$, can be calculated from $\Psi(|r_e - r_o|) = -\alpha_1|r_e - r_o|^{-\beta_1} + \alpha_2|r_e - r_o|^{-\beta_2} = 0$, which implies

$$|r_e - r_o| = \left(\frac{\alpha_2}{\alpha_1}\right)^{\frac{1}{-\beta_1+\beta_2}}. \tag{A.23}$$

Inserting Eq. A.23 into Eq. A.22 yields a restriction for stability

$$\frac{\beta_2}{\beta_1} > 1. \tag{A.24}$$

A.3 Long-Range Instabilities and Interaction Truncation

Let us define the separation distance between the origin ($r_o = 0$) and a particle (located at r) that interacts with a source at the origin. Let us further define a perturbed state (away from r) via

$$\tilde{r} = r + \delta r, \tag{A.25}$$

leading to

$$m\ddot{\tilde{r}} = \Psi(\tilde{r}) \tag{A.26}$$

where r is the perturbation-free position of the particle, governed by

$$m\ddot{r} = \Psi(r). \tag{A.27}$$

Subtracting Eq. A.26 from Eq. A.27, we have

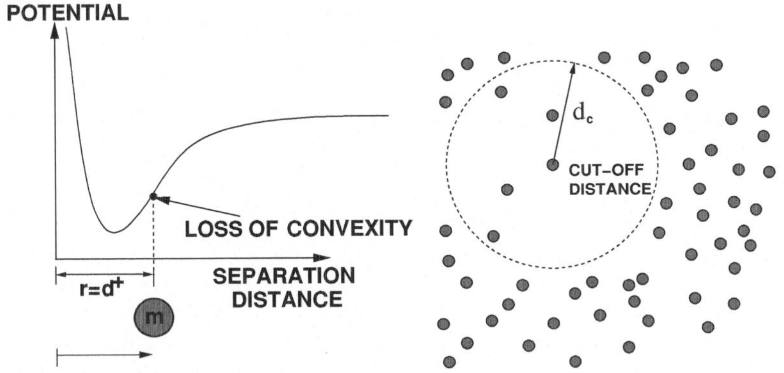

Fig. A.1 *Left* Identification of an inflection point (loss of convexity (Zohdi [195])) and *Right* Introduction of a cut-off function

$$m\delta\ddot{\tilde{r}} = \Psi(\tilde{r}) - \Psi(r) \approx \frac{\partial \Psi}{\partial r}|_{\tilde{r}=r}\delta r + ..., \tag{A.28}$$

resulting in

$$m\delta\ddot{\tilde{r}} \approx \frac{\partial \Psi}{\partial r}|_{\tilde{r}=r}\delta r \Rightarrow m\delta\ddot{\tilde{r}} - \frac{\partial \Psi}{\partial r}|_{\tilde{r}=r}\delta r \approx 0. \tag{A.29}$$

If $\frac{\partial \Psi(r)}{\partial r}$ is positive, there will be exponential growth of the perturbation, while if $\frac{\partial \Psi(r)}{\partial r}$ is negative, there will be oscillatory behavior of the perturbation. Thus, since

$$-\frac{\partial^2 V}{\partial r^2} = \frac{\partial \Psi}{\partial r}, \tag{A.30}$$

we have

$$m\delta\ddot{\tilde{r}} + \frac{\partial^2 V}{\partial r^2}|_{\tilde{r}=r}\delta r \approx 0. \tag{A.31}$$

Thus, for stability, the potential should be convex about r. Clearly, the point at which the potential changes from a convex to a concave character is the point of long range instability (Fig. A.1). As mentioned before, for the central force potential form chosen in this work, it suffices to study the motion in the normal direction, i.e., the line connecting the centers of the particles. For disturbances in directions orthogonal to the normal direction, the potential is neutrally stable, i.e., the Hessian's determinant is zero, thus indicating that the potential does not change for such perturbations. For motion in the normal direction, we have

$$\frac{\partial^2 V}{\partial r^2} = -\beta_1\alpha_1|r - r_o|^{-\beta_1-1} + \beta_2\alpha_2|r - r_o|^{-\beta_2-1} = 0, \tag{A.32}$$

thus leading to

$$|r - r_o| = \left(\frac{\beta_2\alpha_2}{\beta_1\alpha_1}\right)^{\frac{1}{-\beta_1+\beta_2}} \stackrel{\text{def}}{=} d^{(+)}. \tag{A.33}$$

Thus, the preceding analysis indicates that, for the three-dimensional case, an interaction "cut-off" distance (d_c) should be introduced (Fig. A.1)

$$||\boldsymbol{r}_i - \boldsymbol{r}_j|| \stackrel{\text{def}}{=} d_c \leq d^{(+)} \tag{A.34}$$

to avoid long-range (central-force) instabilities.

Remark By introducing a "cut-off" distance, one can circumvent a loss-of-convexity type instability. However, introducing such a cut-off can induce another type of instability. Specifically, if the particles are in static equilibrium, or are not approaching one another, and if the "cut-off" distance, d_c, is much smaller than the static equilibrium separation distance, d_e, then the particles will not interact at all. Thus, we have the following two-sided bounds on the cut-off for near-field forces to play a physically realistic role

$$\left(\frac{\alpha_2}{\alpha_1}\right)^{\frac{1}{-\beta_1+\beta_2}} = d^{(-)} \leq d_c \leq d^{(+)} = \left(\frac{\beta_2\alpha_2}{\beta_1\alpha_1}\right)^{\frac{1}{-\beta_1+\beta_2}}. \tag{A.35}$$

Clearly, since $\beta_2 > \beta_1, d^{(-)}$ is a lower bound (dictated by minimum interaction distance), while $d^{(+)}$ is an upper bound (dictated by (convexity-type) stability).

A.4 Interaction "Strength"

Some important qualitative information can be determined about the interaction law in Eq. 1.51 if we consider a linearization of a single nonlinear differential equation, describing the attraction and repulsion from the origin ($r_o = 0$) of the form[1]

$$m\ddot{r} = \Psi(r), \tag{A.36}$$

where

$$\Psi(r) = -\alpha_1 r^{-\beta_1} + \alpha_2 r^{-\beta_2}. \tag{A.37}$$

Upon linearization of the nonlinear interaction relation about a point r_*,

$$\Psi(r) \approx \Psi(r_*) + \frac{\partial\Psi}{\partial r}\Big|_{r=r_*}(r - r_*) + \mathcal{O}(r - r_*)^2, \tag{A.38}$$

and normalizing the equations, we obtain

[1] The unit normal has been taken into account, thus the presence of a change in sign.

$$\ddot{r} + (\omega_n^*)^2 r = \frac{f^*(t)}{m}, \tag{A.39}$$

where

$$\omega_n^* = \sqrt{\frac{-\frac{\partial \Psi}{\partial r}|_{r=r_*}}{m}}, \tag{A.40}$$

and where

$$f^*(t) = \Psi(r_*) - \frac{\partial \Psi}{\partial r}|_{r=r_*} r_*. \tag{A.41}$$

For the specific interaction form chosen, we have (with $\alpha = \overline{\alpha} m^2$)

$$\omega_n^* = \sqrt{\frac{-\alpha_1 \beta_1 r_*^{-\beta_1 - 1} + \alpha_2 \beta_2 r_*^{-\beta_2 - 1}}{m}} = \sqrt{-\overline{\alpha}_1 m \beta_1 r_*^{-\beta_1 - 1} + \overline{\alpha}_2 m \beta_2 r_*^{-\beta_2 - 1}}, \tag{A.42}$$

and where the "loading" is

$$f^*(t) = -\alpha_1 r_*^{-\beta_1} + \alpha_2 r_*^{-\beta_2} - \alpha_1 \beta_1 r_*^{-\beta_1 - 1} + \alpha_2 \beta_2 r_*^{-\beta_2 - 1}. \tag{A.43}$$

We note that if the following specific choice of parameters is made $(\beta_1, \beta_2) = (1, 2)$, and r_* is chosen as the static equilibrium point, r_e, given by Eq. A.23, then

$$r_* = r_e = \frac{\alpha_2}{\alpha_1}, \tag{A.44}$$

and

$$\omega_n^* = \sqrt{\frac{\alpha_1 \left(\frac{\alpha_1}{\alpha_2}\right)^2}{m}} = \sqrt{\frac{\alpha_1}{m}\left(\frac{\overline{\alpha}_1}{\overline{\alpha}_2}\right)^2} \stackrel{\text{def}}{=} \sqrt{\frac{k^*}{m}}, \tag{A.45}$$

where

$$k^* \stackrel{\text{def}}{=} \alpha_1 \left(\frac{\overline{\alpha}_1}{\overline{\alpha}_2}\right)^2. \tag{A.46}$$

Thus, if we keep the ratio $\frac{\overline{\alpha}_1}{\overline{\alpha}_2}$ constant, while increasing α_1 (while keeping m constant), we effectively increase the interaction/bond "stiffnesses" between particles. Clearly, under certain conditions, a particulate system may "pulse" (oscillate) depending on the character of the interaction. We remark that increasingly smaller ω_n^* indicates that the system tends toward a "regular" (near-field free) particulate system. Smaller ω_n^* would occur with heavier particles or smaller attractive forces.

Index

T. I. Zohdi, *Dynamics of Charged Particulate Systems*, SpringerBriefs in Applied Sciences and Technology, DOI: 10.1007/978-3-642-28519-6,